AF303774

Samurai Swords for the Material Battle

Gendaito of the Taisho and Early Showa Period (1912 – 1945)

Otto Maxein

Bibliografische Information der Deutschen Nationalbibliothek:
Die Deutsche Nationalbibliothek verzeichnet diese Publikation in der Deutschen Nationalbibliografie; detaillierte bibliografische Daten sind im Internet
über http://dnb.dnb.de abrufbar.

Herstellung und Verlag: BoD – Books on Demand, Norderstedt

ISBN 978-3-7534-7141-9

Foreword

Japanese swords have an excellent reputation among connoisseurs. The complex processes during forging or the individualization of a cutting edge through the application of an aesthetic hamon fascinate far beyond the borders of Japan. To this day, the samurai sword is so deeply rooted in Japanese culture that it is inseparable from it. The Japanese swordsmiths have always played a major role in this.

But what drove soldiers, long after the era of the samurai, to go into the material battles of the Second World War with the drawn sword? In a conflict that was fought with means that made a fight with the blade not only hopeless, but downright suicidal. Why was it important for pilots to take their swords on board their planes, even though there was no practical use for them there? What was inherent in these blades that they exerted such fascination and effect on their bearers? What makes the swords forged at Yasukuni Shrine or the swords of the two generations of Minamoto Yoshichika special? How do they clearly stand out from the mass of swords forged during the Taisho and early Showa period? What makes them swords worth preserving? And what is the truth of the saying "The sword is the soul of the samurai"?

The author explores these questions in his book. He deals with the subject with great attention to detail. He explains, without glorifying, what makes the fascination of these weapons and takes the reader with him into the material battles of the Second World War, in which the samurai sword played a more important role for the imperial soldiers than one would assume at first glance.

Oliver Gerigk

Index of Contents

In Grateful Memory of My Father
Otto Rudi Maxein
(* 11.06.1917, † 06.05.1990)

I dedicate this book to my father, who survived the Second World War as a front-line soldier from the Polish campaign to his capture at the end of the war. As a tank hunter, he took part in several major tank battles in the East, was wounded and was awarded the medal Winter Battle in the East 1941/42, called "Frozen Meat Medal" by the fighting soldiers, the Iron Cross and the Bronze Close Combat Clasp, which means that he has participated in at least 15 hand-to-hand fights. It was only now, long after his death, through reading Erich Maria Remarque, Edlef Köppen and Andreas Engermann that I realized what almost unimaginable hardships he must have endured, what brutal dehumanization he must have experienced in hand-to-hand combat man-to-man.

Like all soldiers of the fighting troops, he went through hell and experienced the horror of war in his own body and soul. Nevertheless, I remember my father as a life-affirming person who, after his release from war captivity, looked ahead like hundreds of thousands of other war repatriates and made his contribution to the reconstruction and economic miracle of the young Federal Republic of Germany. To him and my dear mother I owe my sheltered, happy childhood and carefree youth.

My father accompanied my wild years with indulgence and understanding. Friends were the most important thing to me during this time. So when I once called him my friend, he answered with a smile: "I am not your friend, I am your father." Today I know what he wanted to tell me. Many a "friend" has disappointed me in life – my father never did! I have learned a lot for life from him and much of what makes me what I am today I owe to him. His upright character and his moral courage are still my role model and orientation today. I am proud of my father and love him – beyond death. Unfortunately I never told him that. I hope he knew it.

Otto Willi Maxein

Special Thanks

No reference book can do without valid sources and extensive research. This is especially true for topics that have not yet been conclusively covered in the relevant literature. Here, the support of third parties is needed, who help to close gaps in knowledge or who provide visual material to be integrated into the work. I would therefore like to take this opportunity to thank all supporters who have given me advice and support and made this book possible in its present form!

My special thanks therefore go to

Annegret Wieland M.A.
Department of Culture and Public Relations
Embassy of Japan, Berlin, Germany

David Ito
Aikido Chief Instructor
Aikido Center Los Angeles, USA

Kazushige Tsuruta
Japanese Sword Shop Aoi-Art, Tokyo, Japan

Prof. Dr. Diethelm Düsterloh
Paderborn, Germany

Oliver Gerigk
Bochum, Germany

Christoph Pieper
ibs Sicherheitstechnik, Gelsenkirchen, Germany

Martin Strebel,
Juwelen und Asienkunst, Wiesbaden, Germany

Paul Underwood
English Training, Dortmund, Germany

So Shihan Wolfgang Wimmer, Kyoshi, Hachidan
Repräsentant Dai Nippon Butoku Kai
Ehrenpräsident Verband asiatischer Kampfkünste e.V.
Meitingen, Deutschland

Last but not least, my special thanks go to my dear wife, who has always been sympathetic to my passion for samurai swords and has endured to the limits of patience my frequent mental absences while working on this book and the weeks-long conversion of our living room into a photo studio.

Otto Maxein, "A lucky day mid-autumn 2020"

Syntax, Formatting

A special feature of the German text edition with regard to Japanese nomenclature is the use of upper and lower case letters, since in the Japanese script, as in other written languages (Hebrew, Arabic, Chinese, etc.), the distinction between upper and lower case letters is unknown.

It was simpler in the English text output, because in English, too, lower case is used at first except at the beginning of the sentence. Exceptions apply e.g. to proper names, salutations, geographical designations or in headings.

In accordance with the English text edition, terms of the Japanese nomenclature are also written in lower case in the German text edition, except at the beginning of the sentence. Exceptions are terms that are capitalized in accordance with German spelling, insofar as this appears to make sense in the context due to the way the language is perceived.

In translations, names and terms related to Japanese sword terminology have been adopted according to the spelling of the sources for the sake of translation accuracy.

Words printed in italics in the continuous text are either proper nouns or serve to emphasize certain terms or originate from Japanese nomenclature. Explanations can be found in the glossary. Furthermore, captions are printed in italics to distinguish them from the body text.

Image Material, Own Photos

Photos on pages 22, 41, 42, 43 and 44 courtesy of Kazushige Tsuruta-San, Aoi-Art, Tokyo, Japan. The "Japanese Sword Shop Aoi-Art" and the "Japanese Sword Online Museum" are always worth a visit to the website of Kazushige Tsuruta-San. The remaining images are either from press or state archives and are expressly marked "public domain" or are in the possession of the author with corresponding rights of use or as originals.

A sometimes severely reduced image quality is due either to natural aging of the image material used or to the resolution of some image files being too low for printing. In view of the book title, these images were nevertheless included because they seemed suitable for visualizing the topic and have lost none of their expressiveness despite the reduced image quality.

Lovers of Japanese blades know that their observation requires the right light. For example, when examining the tempering line, we need a specific light source to guide the blade along. A photograph, on the other hand, is a snapshot and always captures a blade in the light of a single moment. I photographed my swords shown in this book myself. Thus I spent many days - my wife claims weeks - patiently experimenting and learning to live with compromises. The photos therefore give only a very distant impression of the beauty of the blades and their inherent activities. But anyone who knows Japanese swords knows anyway that they must be taken in hand and studied in the right light for them to reveal themselves to us in all their beauty and perfection.

Samurai Swords for the Material Battle

Gendaito of the Taisho and Early Showa Period (1912 – 1945)

In no other culture has a weapon in the course of its history attained such a high spiritual and social significance as the samurai sword in Japan. For more than a thousand years it was not only considered the soul of the samurai, but as a feared weapon of an elitist and death-defying warrior caste it was the all-controlling status symbol, which the Japanese paid awe, respect and almost religious submission to. Triggered by the *Meiji Restoration* and Japan's path to modernity, social change and the striving for progress at the end of the 19th century sealed the fate of the samurai and their swords, which henceforth seemed like anachronisms and relics of a past epoch.

But although samurai swords, like infantry rapiers or cavalry sabres in Europe, had lost their importance as weapons of war with the introduction of machine guns and tanks, the belief in the spiritual power of the samurai sword was so deeply rooted in Japanese thinking that after their initial exile at the beginning of the *Meiji Restoration*, these swords experienced a military renaissance during the *Taisho* and *early Showa period*. In addition to the countless machine blades that were made in Japan for parade purposes or for lower ranks, the demand of Japanese officers for new swords traditionally forged in the spirit of *bushido* revived an ancient craft.

Experienced swordsmiths were activated throughout Japan, training swordsmiths in various forging centers and forging their own swords. Among the most famous places of activity were the *Minatogawa Shrine*[1] in Kobe on the banks of the Minatogawa (Minato River), where swords were forged for the Imperial Navy, and the *Yasukuni Shrine*[2] in Tokyo, where enormous efforts were made to revive the spirit of the samurai by returning to traditional forging methods in the *Yasukuni-to*. There were also other focal points for the mass production of army swords, such as the city of Seki in Gifu Prefecture. However, it should be noted that the vast majority of swords produced in Seki were sufficient as weapons, but could not meet the high standards of the Japanese sword as an art object.

This is true for swords of all epochs. Not every samurai sword that suitable as a weapon was artistically valuable, although it is often the case that the artistically perfect blade is also the better one for practical use.[3] It is also true that the gunto, which were traditionally forged by Japanese swordsmiths until the Second World War, were probably the last samurai swords with which Japanese soldiers went into battle for the emperor and the empire and, as once like the samurai, serving fulfilled their duty. Thus the gunto of the Taisho and early Showa period mark a historical turning point in the history of samurai swords. The blades of Yasuoki and the two generations of Minamoto Yoshichika, forged during this period and presented in this book, bear witness to the undisputed mastery of these smiths and their high claim to provide Japan's "last samurai" with swords that were not mere weapons, but ideally combined weapon and work of art in one and the same blade.

---

[1] http://de.wikipedia.org/wiki/Minatogawa-Schrein

[2] http://de.wikipedia.org/wiki/Yasukuni-Schrein

[3] Hagenbusch, Michael, Beitrag im Katalog zum Ersten Europäischen Symposium „Die Kunst der Samurai", Deutsches Klingen-Museum, Solingen 1984

源義親

Minamoto Yoshichika
Shodai and Nidai

Minamoto Yoshichika is counted among the few important swordsmiths of the Taisho and Showa Period. His swords were known for their exceptional sharpness and combat capability and were worn by the Imperial Guard and famous swordsmen. At the coronation of Emperor Hirohito in 1928, he was one of a small group of swordsmiths who were selected among the best in Japan to forge the swords that, following an old tradition, were presented to high dignitaries during the coronation ceremonies. Nevertheless, there are occasional attempts to relativize the significance of his work by the objection that he is said to have forged swords from *western steel.*

This is to imply that swords of Minamoto Yoshichika are not genuine *nihonto* or *gendaito* because they were not forged from *tamahagane.* If this were true, Emperor Hirohito, who certainly had more sword education than many a self-proclaimed expert in the western hemisphere, would have lost face the moment he called Minamoto Yoshichika to Tokyo and to become the Imperial Swordsmith. In addition there are authorities in the motherland of the Japanese sword which decide competently whether a traditionally forged sword also fulfills the artistic demands of a Japanese sword and is thus recognized as a sword worth preserving or not. These are primarily the two major Japanese sword societies *Nihon Bijutsu Token Hozon Kyokai (NBTHK)* and *Nihon Token Hozon Kai (NTHK).*

Minamoto Yoshichikas swords have received expertises *(origami)* from the *NTHK* as well as from the *NBTHK* and *Fujishiro* and are recognized by the leading sword authorities of Japan without any doubt as genuine gendaito.[4] In TOKO TAIKAN Minamoto Yoshichika is listed on page 758 under YOS1067, his swords are classified as *"Highest Grade Gendaito"*.[5] In addition to the high distinction Minamoto Yoshichika received through his appointment as Imperial Swordsmith, his position as an eminent swordsmith is further evidenced by the fact that he is the only swordsmith of the Taisho period to have been included in *Fujishiro's* authoritative encyclopedia *"Nihon Toko Jiten, Shinto-hen"*.[6]

Although Minamoto Yoshichika has gone his own way to produce powerful sword blades with special cutting ability[7], he has used the traditional techniques of Japanese swordsmiths *(katana-kaji)* to create high quality gendaito[8], that are sought after by collectors today.[9] Not least because it is now known that blades from Minamoto Yoshichika can compete even with high quality swords from the *Kamakura* and *Muromachi period.*[10]

But before we go back to the objection mentioned at the beginning which is ultimately aimed at the dispute between *"tamahagane"* and *"western steel"*, let us talk about the swordsmith and his work, as far as research on the life and work of Minamoto Yoshichika has revealed.

---

[4] Stein, Richard, Japanese Sword Guide,
http://www.japaneseswordindex.com/yoshchik.htm
[5] Tokuno, Kazuo, TOKO TAIKAN, YOS1067, 2004
[6] Fujishiro, Yoshio, Nihon Toko Jiten, Shinto-hen, Tokyo 1961
[7] Slough, John Scott, An Oshigata Book of Modern Japanese Swordsmiths 1868 – 1945, Rivanna River Company, 2001
[8] http://www.worthpoint.com/worthopedia/rare-mint-antique-japanese-samurai-katana-sword
[9] Couch, Paul and Matsuoka, Yumiko, Thoughts on Nihonto – Gendaito, ISF-AL/GA Newsletter, May 2002, Vol. 3, Issue 3
[10] http://www.worthpoint.com/worthopedia/rare-mint-antique-japanese-samurai-katana-sword

Minamoto Yoshichika, Shodai

Minamoto Yoshichika came from *Shibamishima, Japan*. His civil name was *Mori Hisasuke*.[11] A different spelling can be found in John Scott Slough's book *"An Oshigata Book of Modern Japanese Swordsmiths 1868 – 1945"*. Here Minamoto Yoshichika is listed on page 196. His swords are classified as *"High Grade Gendaito"* and their value is estimated at 1.5 million Japanese Yen. Alternatively we find here the spelling *Mori Kyusuke* for his civil name.[12]

Minamoto Yoshichika is said to have called himself the last descendant of Sanjo Munechika from Yamashiro (ca. 987).[13] Even if one comes across this reference again and again, researches for this could so far not produce a secured proof. Despite this, Minamoto Yoshichika is one of the few swordsmiths of the Taisho and Showa period who is considered one of the most important swordsmiths of his time by the leading authorities and experts on the subject.[14]

During the Meiji and Taisho period, Minamoto Yoshichika worked in Musashi Province. In 1926 he followed the call of Emperor Hirohito to Tokyo, where he forged swords as an imperial artisan by order of the emperor.[15] Most of his famous works come from the Taisho and early Showa period. He preferred to work in the *Bizen-den-gunome-choji style* and his blades are known for their exceptional sharpness and cutting ability.

---

[11] Stein, Richard, Japanese Sword Guide,
http://www.japaneseswordindex.com/yoshchik.htm
[12] Slough, John Scott, An Oshigata Book of Modern Japanese Swordsmiths 1868 – 1945, Rivanna River Company, 2001
[13] Stein, Richard, Japanese Sword Guide, siehe Referenz-Nr. 11
[14] JSS newsletter, in extracts published on
http://www.nihonto.com.au/html/minamoto_yoshichika_tachi.html
[15] http://www.samuraisam.net/tachiofyoshichika.html

Minamoto Yoshichika forged swords for the court and the Imperial Guard. But also famous martial artists and swordsmen of the early twentieth century preferred his blades. So, one of the swords preferred by *Hakudo Nakayma*[16] came from the forge of Minamoto Yoshichika. *Hakudo Nakayma* (other spelling *Hiromichi*) was the most famous sword fighting master of his time and is one of only ten *Budo Grand Masters* who were awarded the title *Meijin ("Brilliant Man")*,[17] the highest distinction in Budo. At the Toyama Military Academy, *Hakudo Nakayma* was a member of the commission that created the curriculum for sword fighting. He was also the official sword fighting instructor of the Japanese Navy and Imperial Guard. *Hakudo Nakayma* was born in Kanazawa, Ishikawa Prefecture on February 11, 1873, and died on December 14, 1958, at the age of 85. He found his final resting place at Tenshin Temple in Minato District, Tokyo.

---

[16] http://de.wikipedia.org/wiki/Nakayama_Hakudo
[17] http://de.wikipedia.org/wiki/Meijin

A rare contemporary document: the picture from the photo album of a Japanese soldier shows Hakudo Nakayma, Budo Grand Master and sword fighting instructor of the Imperial Guard, together with members of the Imperial Guard.

Minamoto Yoshichika's clients knew that he forged sharp and highly resilient swords for the toughest hand-to-hand combat man-to-man for life and death. They differ clearly from the long and wide blades with spectacular *hamon*, as we often find today among Iaido students. Instead, his work resembles slim *koto* blades with functional *hamon*. Despite their low weight they convince with high torsional strength, perfect balance and impressive handling, but more about this later.

A certain number of swords forged by Minamoto Yoshichika have been subjected to a cutting test *(tameshigiri)* by *Hakudo Nakayama*. On swords that have been tested by him, we find the stamp *"Hakudo Tameshigiri Sho"* on the tang.[18] Most

---

[18] Stein, Richard, Japanese Sword Guide,
http://www.japaneseswordindex.com/yoshchik.htm

swords of the Imperial Guard are forged by Minamoto Yoshi-chika and tested by *Hakudo Nakayama*. The following incident is handed down:

Obata Toshishiro, a well-known author, practicing swordsman and expert on the subject, reports that he found a noteworthy text in his Willis Hawley collection, from which it is clear that *Hakudo Nakayama* tested swords for the Imperial Guard on pigs, with the blades cutting smoothly through the bodies. It is said that he demonstratively cut through the hip bones of a pig with a sword to prove to Vice Admiral Oyamada how a really sharp sword should cut.

When the admiral inquired about the swordsmith, *Hakudo Nakayama* explained that the sword came from the forge of Minamoto Yoshichika. Thereupon Admiral Oyamada issued a decree that all swords of the Imperial Guard must be forged by Minamoto Yoshichika from now on and must be undergo a cutting test seven times by *Hakudo Nakayama*. *Obata Toshishiro* states that a total of 490 of these swords were tested and accepted for the Imperial Guard by *Hakudo Nakayama*.[19]

This number alone, only 490 swords of Minamoto Yoshichika were tested and accepted by Hakudo Nakayama for the Imperial Guard, makes clear why these swords are extremely rare and are sought after by serious collectors worldwide. This becomes even more impressive when one compares this small number of swords with the well over one million Japanese swords that were brought to the United States by American soldiers after the war.[20] John M. Yumoto's estimate of 250,000 to 350,000 swords that were brought to the United States does not change much. Even so, the hit rate to find another blade of Minamoto

---

[19] http://www.nihonto.com.au/html/minamoto_yoshichika_tachi.html
[20] Sly, Christopher, More Thoughts On Gendaito, Sept. 1992, updated by Bowen, Chris and Massey, Denny, March 2015, http://www.nihontocraft.com/Gendaito.html

Yoshichika would be only 1.4 ‰ (per thousand).[21] In fact this chance should be much lower, since these blades, as far as they still exist, have long since passed into private ownership. This thesis is further supported by Obata Toshihiro's remark in his article *"Swords and Tradition"*, according to which he personally knows only three swords by Minamoto Yoshichika with a cutting test by Hakudo Nakayama, which appeared in the USA.[22]

In view of such numbers it becomes clear how difficult it is meanwhile to find a Minamoto Yoshichika with cutting test. It might be even more difficult to find a blade signed with Nidai Minamoto Yoshichika, because swords signed with Nidai Minamoto Yoshichika are simply even rarer.

---

[21] Yumoto, John M., Das Samuraischwert, Ein Handbuch, Ordonanz-Verlag Strebel GmbH, Wiesbaden, 2004
[22] Obata Toshishiro, Swords and Tradition, https://kenjutsu-ryu.livejournal.com/29096.html

Minamoto Yoshichika, Nidai

It is considered certain that there was a second generation Minamoto Yoshichika, Nidai Minamoto Yoshichika. Nidai Minamoto Yoshichika was the natural son of Minamoto Yoshichika and worked with his father during the Taisho and Showa period. Nidai's style is similar to that of the Shodai, but the son forged more expressive blades. The *hamon* is more active with countless *ashi* and *yo* in *ko-nie*, while the *hada* is predominantly dense *masame*. Blades of the Nidai show *utsuri*.

Although the aesthetics of his blades inspire enthusiasm and the quality of his work leaves no doubt as to his mastery, the life and work of Nidai has remained largely hidden from the public until today. Even in the relevant literature there is little to read about it. One reason for this may be that works by Nidai are rare and in the meantime are hardly offered at all. *Richard Stein* writes about this on his renowned and recommendable website *"Japanese Sword Guide"*: "Examples of the work of the second generation with the inscription "Nidai" at the beginning of the signature are rare." Stein further emphasizes: "The works of the two generations Minamoto Yoshichika must not be confused with swords from the *Showa Era*, which are signed *"Noshu Seki ju Yoshichika"*. These smiths belong to the insignificant swordsmiths."[23] An unknown number of swords signed "Minamoto Yoshichika" are said to have been forged together by father and son. This practice was not uncommon, as outstanding swords were often forged in collaboration between two masters. But this could have meant that a part of the son's work was absorbed by the father's work. After the death of his father, Nidai Minamoto Yoshichika is said to have given up forging forever. This could be another reason why blades of the Nidai are extremely rare.

---

[23] Stein, Richard, Japanese Sword Guide,
http://www.japaneseswordindex.com/yoshchik.htm

How sought-after swords of the Nidai are today and how diffi-cult it is to buy another Nidai Minamoto Yoshichika is also shown by the fact that fakes have already appeared on the market, which is rather unusual for *gendaito*. The author is also familiar with two attempts at forgery due to his many years of involvement with the subject. In the first case it was an original Minamoto Yoshichika, which was later preceded by a "Nidai". The different style of the original signature revealed the forgery attempt. In the second case, the blade alone did not meet the quality and artistic demands of a Minamoto Yoshichika. The examination of the tang *(nakago)*, which deviated blatantly from the aesthetics of an original work, as well as a signature that did not correspond to the style of the original signature, ultimately confirmed the suspicion that had already arisen during the examination of the blade.

Nidai Minamoto Yoshichika is said to have worked for the army as *Rikugun Jumei Tosho* and was therefore supplied with *tamahagane*. This thesis is supported by the fact that during his research for this book, the author came across blades in the WWW three times, which were signed with *"Nidai Minamoto Yoshichika Saku Kore"* and which additionally carried the *"Kikusui-mon"* on the tang. The *Kikusui-mon* is the crest of Minatogawa Shrine and depicts a chrysanthemum on the waves of the Minatogawa River. One of these swords is illustrated and described in detail in the "Japanese Sword Online Museum" of the company "Aoi-Art", Tokyo, Japan.[24]

The fact that Nidai blades exist that carry a *Kikusui-mon* on the tang proves that Nidai Minamoto Yoshichika also worked at the Minatogawa Shrine and forged swords for the Imperial Japanese Navy there. At the instigation of the Navy, traditional samurai swords were forged at the shrine from 1941 by re-nowned swordsmiths for graduates of the Naval Academy and

---

[24] https://www.aoijapan.net/katana-minamoto-yoshichika-saku-kore/

high-ranking naval officers. In addition to the signature of the swordsmith, these swords also bear the *Kikusui-mon* on the tang.[25] It is said that only a few hundred swords were forged at the Minatogawa Shrine, which is why they are sought-after by collectors and are just as sought-after today as the swords forged at Yasukuni Shrine.[26]

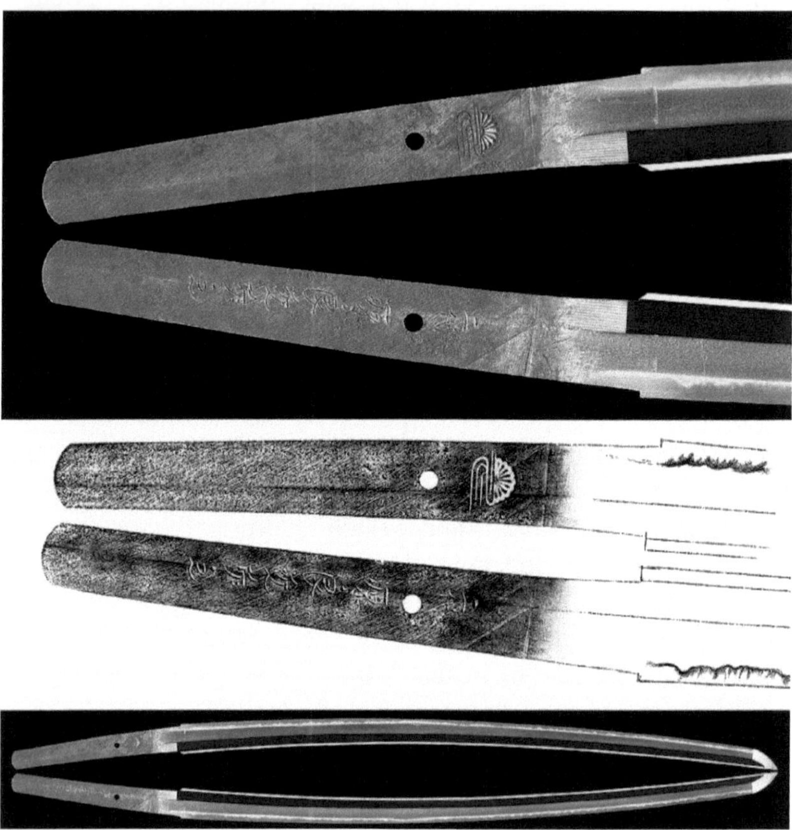

A cultural-historically important work of the second generation Minamoto Yoshichika. Tachi-mei "Nidai Minamoto Yoshichika

---

[25] Wallinga, Herman A., Gendaito Made at the Minatogawa Shrine, Verlag Herman A. Wallinga, 2000
[26] https://de.wikipedia.org/wiki/Minatogawa-Schrein

Minatogawa Shrine is a Shinto shrine in Kobe, Japan, built to worship Kusunoki Masashige, who committed seppuku after losing the Battle of Minatogawa. Imperial forces led by Kusunoki Masashige attempted to intercept the rebels led by Ashikaga Takauji on July 5, 1336. In the process, Emperor Go-Daigo decided against the strategy of his commander, which ultimately led to the defeat of the imperial forces. Despite this defeat, Kusunoki Masashige is revered at the Minatogawa Shrine to this day for the loyalty he demonstrated to the emperor. His wife is enshrined in a side shrine.[28]

---

[27] https://www.aoijapan.net/katana-minamoto-yoshichika-saku-kore/
[28] https://de.wikipedia.org/wiki/Schlacht_am_Minatogawa

Picture page 23: The Minatogawa shrine on an old Japanese postcard. In focus a 28 cm howitzer L/10. Developed in 1884 by the British Armstong Company, the gun was used by the Imperial Japanese Army from 1892 to 1945 and was manufactured in Japan at the Osaka Artillery Arsenal. A total of 220 pieces were produced. The gun was mounted on a steel mount with a turntable. Total weight 10.75 tons, barrel length 2.86 m. It took two to four days to get the howitzer into position. An ammunition elevator was used for loading. The projectile

weighed 217 kg and reached a muzzle velocity of 314 m/sec. The maximum range was 7800 m.[29] *The howitzer served in coastal batteries and as a siege gun, including during the Russo-Japanese War in the siege of Port Arthur (19.07.1904 - 20.12.1904). Port Arthur was the focus of heavy fighting because it was the only ice-free deep-water port in the Far East and of strategic importance to the Russian Pacific Fleet.*[30]

Photo page 24: Russo-Japanese War, Port Arthur, Liaoning Province, China: A siege gun takes fire at the Russian Pacific Fleet lying in Port Arthur harbor. Out of the gun smoke, the 28-cm shell flies toward the harbor. On 5 December 1904, after preceding massive barrage, the Japanese overran Russian positions and occupied a 203-meter hill. From here, an artillery observer directed fire on the Russian fleet, systematically sinking one ship after another. The battleship Poltava sank on December 5, followed by the battleship Retvizan on December 7. On December 9, the battleships Pobada and Peresvet and the cruisers Palladan and Bayan were sunk. All six ships were later raised by the Japanese and recommissioned after the war. The 154-day successful siege of Port Arthur is considered the greatest military achievement of General Nogi Maresuke, who committed seppuku in 1912 after the death of Emperor Meiji in the tradition of the samurai to serve his master even in the other world.

---

[29] https://en.wikipedia.org/wiki/28_cm_howitzer_L/10
[30] https://de.wikipedia.org/wiki/Belagerung_von_Port_Arthur

*The artillery observer's view of Port Arthur from elevation 203.
After the shelling, combat ships of the Russian Pacific Fleet lie
aground.*

Coronation Tachi by Minamoto Yoshichika

On the occasion of the coronation celebrations in 1928, 20 swords, which Emperor Hirohito had ordered from important swordsmiths in the country, were presented to the highest dignitaries. Among them was at least one tachi by Minamoto Yoshichika. [31] How many tachi Minamoto Yoshichika actually forged for this occasion on behalf of the court could not be clarified in the course of this research. It is interesting, however, that in recent years three more swords have come onto the market, which prove or rather allow the conclusion to be drawn that these are also tachi forged by Minamoto Yoshichika on the occasion of the coronation commissioned by the court.

Example 1: On October 16, 2012, a sword made by Minamoto Yoshichika, which Minamoto Yoshichika had forged on behalf of the court at the coronation celebrations as a gift to a high-ranking dignitary, was auctioned off at Bonhams, New York. The sword was described for the auction as follows:

"Los 1183

An Imperial Presentation tachi By Minamoto Yoshichika (Mori hisasuke), dated 1928

Sold for US$ 18,750 including surcharge

The small, elegant blade of *hon-zukuri, iori-mune, koshi-zori, ko-kissaki* configuration and forged in very tight *ko-itame* and tempered with a narrow *sugu-ba hamon*, with a *ko-maru boshi*, the tang *ubu, kuri-jiri* with one hole and signed and dated *Minamoto Yoshichika* and dated *Showa san-nen chu-shu* (mid-autumn 1928); one-piece silver *habaki*; 25 3/4in (65.5cm) long; in a *shirasaya*.

---

[31] http://www.samuraisam.net/tachiofyoshichika.html

The *tachi-koshirae* comprising a black-lacquer *saya* decorated with paulownia crests in gold *hiramakie* and mounted with silver hardware, the hangers bearing purple and blue doe-skin straps; the *tsuka* applied with white ray-skin and fitted with "tweezer"-type silver *menuki* carved with paulownia crests; silver *fuchi-kashira*; silver *san-mai awase* silver *tsuba*.

With a 180-page book entitled "The Enthronement of the One Hundred and Twenty-Fourth Emperor of Japan", and a silver coronation medal with gilt *kiku*.

The tachi in box and book along with medal were presented to perhaps fewer than twenty visiting dignitaries attending the coronation of Hirohito."[32]

Example 2: Another coronation *tachi* from Minamoto Yoshichika has been offered and sold by the company "Rice Cracker". The sword was described as follows:

"A very nice complete package in an *efu no tachi* mount, destined to be a gift at the highest level, either from the Imperial Family or another high ranking personality. The mounting in very good condition, all metal parts are heavily gold-plated. A small bundle of rice on the handle is missing, but apart from that, everything is in excellent condition. The scabbard is beautifully finished in high-quality *nashiji* gold lacquer.

...The blade is a rare and unusual sword by Minamoto Yoshichika. Forged in the style of an old *tachi* of the *Heian Period*, it is a very beautiful and elegant blade. Shodai Yoshichika ("Yoshichika the 1st") is known to have forged for the Imperial Palace Guard and many of his blades were tested by the famous Kenshi Nakayama Hakudo. It is believed that Hakudo tested all the blades of Yoshichika that were intended for the Imperial Palace Guard, and he (Hakudo) also carried such a sword. The

---

[32] http://www.bonhams.com/auctions/20503/lot/1183/

blade needs polishing, but you can see the well done *hada (grain structure)* and a very nice *suguha hamon.*

Above the *hamon* you can see a nice well forged *hada* with *ji nie.* The blade is flawless and should be very nice to polish. The shape of the blade is very elegant and looks like a *tachi* of the *Heian Period.* The present sword is rare because it is also dated. Blades from Yoshichika often have no date on the tang. The blade together with the *tachi* stand, which is only slightly damaged and in quite good condition, was given to a high-ranking American officer during the occupation (of Japan) and then brought to the States. Signed *Kin Yoshichika Kore O Saku Showa San Nen Ni gatsu Hi* (1928), blade length *(nagasa)* slightly over 25 1/16" (ca. 63,66 cm)."[33]

The detailed descriptions in both cases clearly show the extraordinary quality and beauty of the blades. The elegance of old *Koto-tachi,* the flawlessly forged dense *hada* and the stylish execution of the *hamon* speak for Minamoto Yoshichika's mastery and artistic skills. In connection with the nearly identical blade lengths these *tachi* appear like twins from the golden age of the Japanese sword and make these swords, due to their artistic execution and their special historical background, worth preserving and cultural-historically important swords of high significance.

Example 3: A *tachi* by Minamoto Yoshichika signed *"Minamoto Yoshichika"* and dated *"Showa san-nen chu-shu"* (mid-autumn 1928).

For the description, the description of the Bonhams-Tachi (example 1) was copied into the text (in quotation marks). The specifications about *bohi* and *sori* was added:

---

[33] http://ricecracker.com/

"The small, elegant blade of *hon-zukuri, iori-mune, koshi-zori, ko-kissaki* configuration and forged in very tight *ko-itame* and tempered with a narrow *sugu-ba hamon*, with a *ko-maru boshi*, the tang *ubu, kuri-jiri* with one hole and signed and dated *Minamoto Yoshichika* and dated *Showa san-nen chu-shu* (mid-autumn 1928)." One-piece gold-plated copper *habaki, nagasa* 63.5 cm, in a *shirasaya*. **Addition:** *bohi hisaki agari, maru dome; sori 23 mm.*

The further description was copied from the description of the Rice Cracker Tachi (example 2):

"The shape of the blade is very elegant and looks like a *tachi* of the *Heian Period*. This sword is rare because it is also dated. Blades from Yoshichika often have no date on the tang."

The similarity of the work to the *tachi* described above (examples 1 and 2) and the rare and special dating "mid-autumn 1928" – in 1928 the enthronement of Emperor Hirohito took place – allow the conclusion that the present blade is another *tachi* forged by Minamoto Yoshichika on behalf of the court on the occasion of the coronation ceremonies of Emperor Hirohito and brought abroad after the end of the war by the confiscation of the victorious powers. The blade has been re-polished. Furthermore a new *habaki* and a new *shirasaya* were ordered and the blade was finally presented to the NTHK for examination in a *shinsa*. The *NTHK Shinteisho* with tang abrasion *(oshigata)* and the personal seals of six appraisers is shown on page 35. Thanks to the collector of that time, this special sword did not perish and was preserved for posterity. The *tachi* is now in the possession of the author.

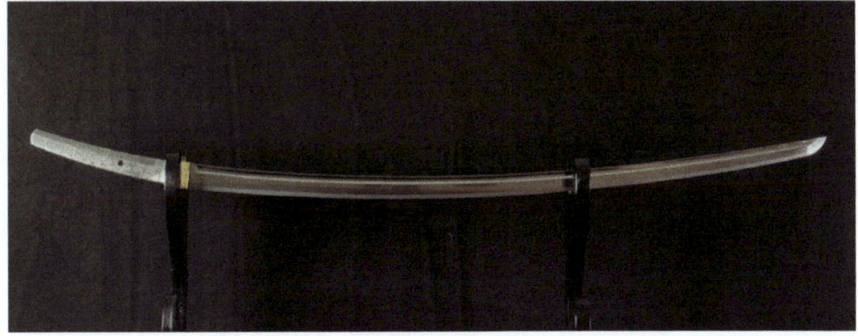

Shodai Minamoto Yoshichika Example 3: The slender, elegant blade with perfectly cut bohi and aesthetically applied hoso suguha hamon in shirasaya.

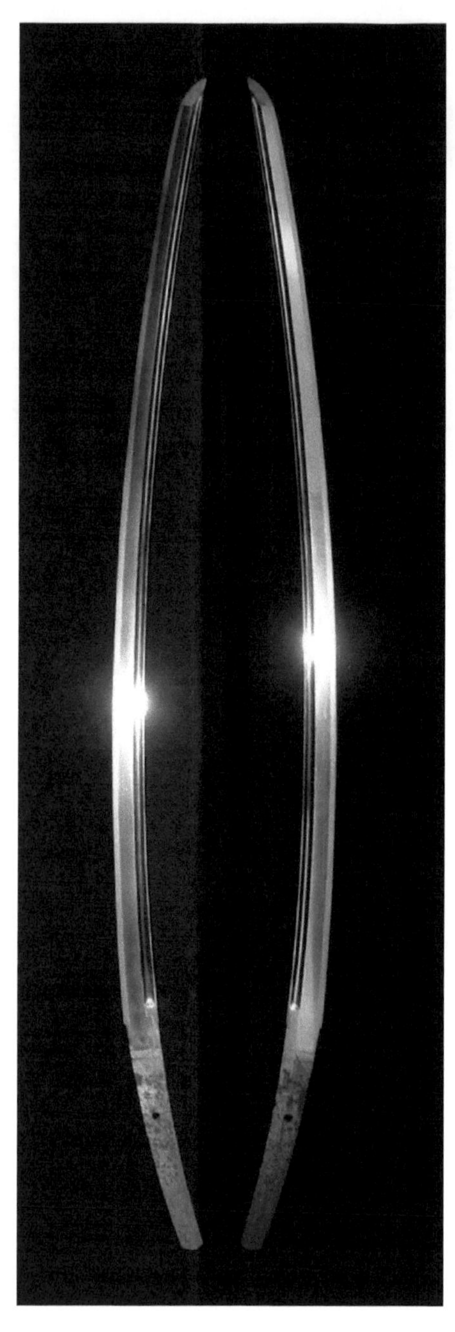

32

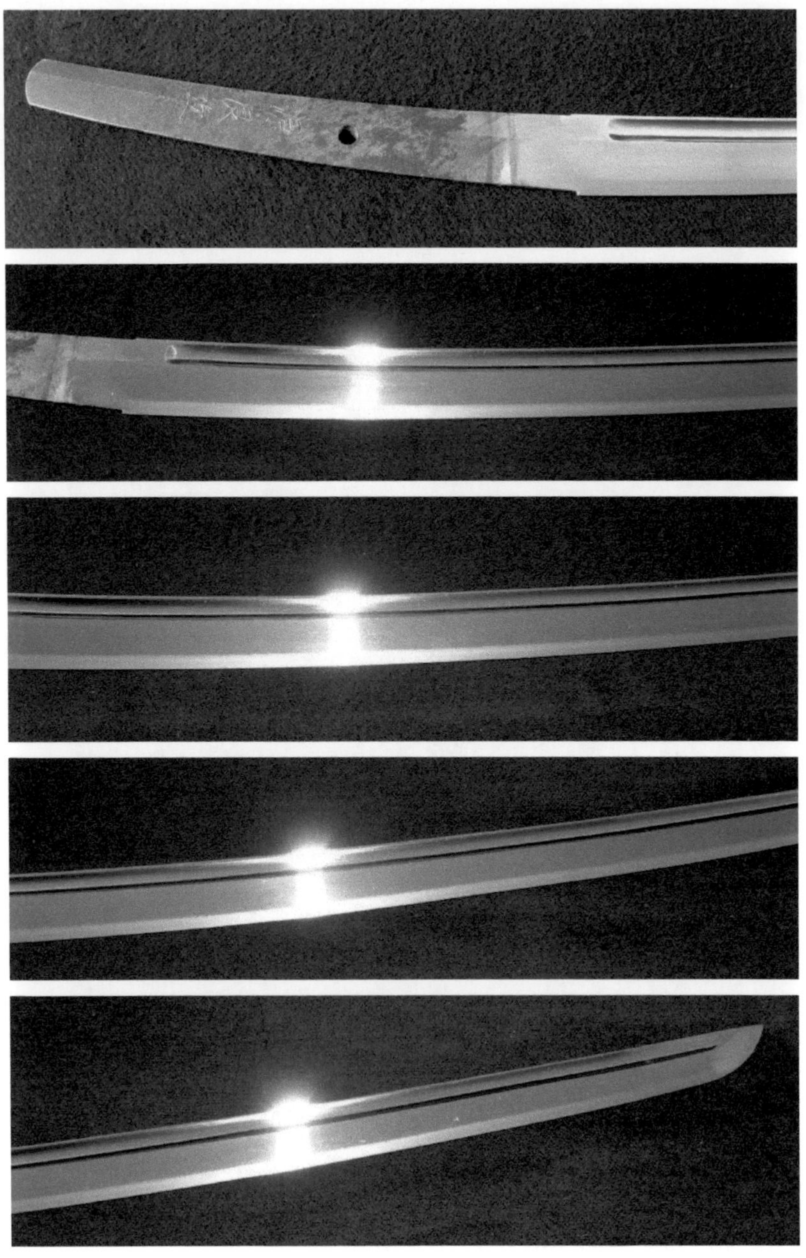

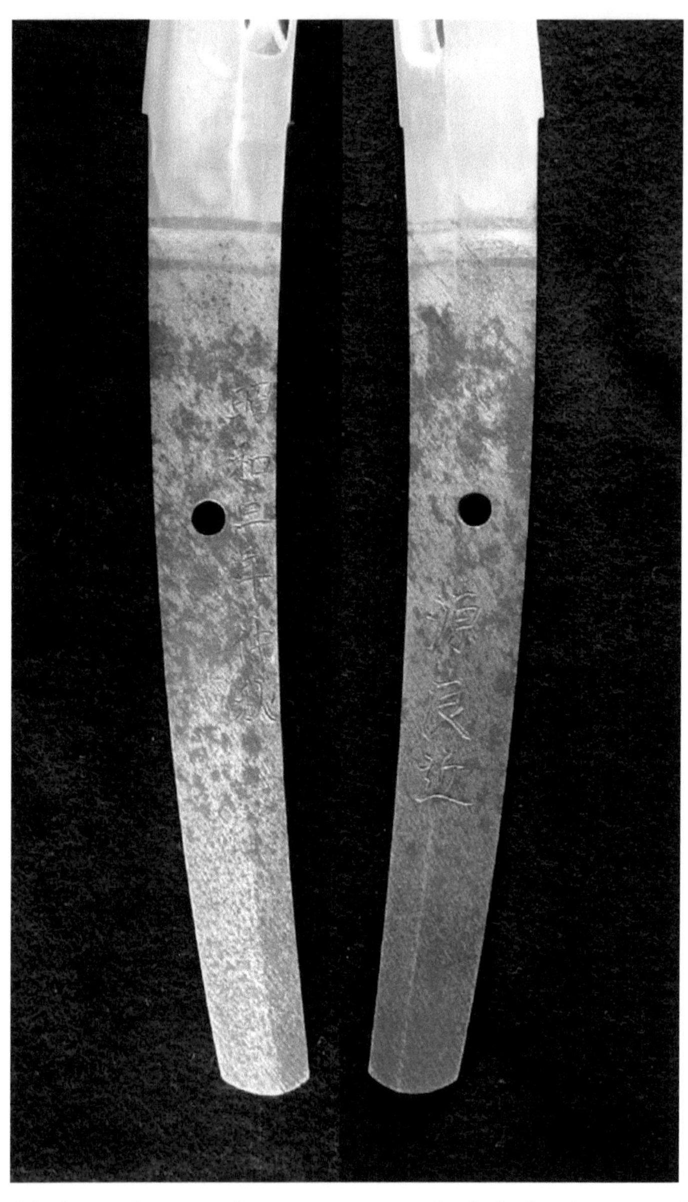

The blade tachi-mei "Minamoto Yoshichika", the ura dated "Showa San-Nen Chu-Shu" (mid-autumn 1928).

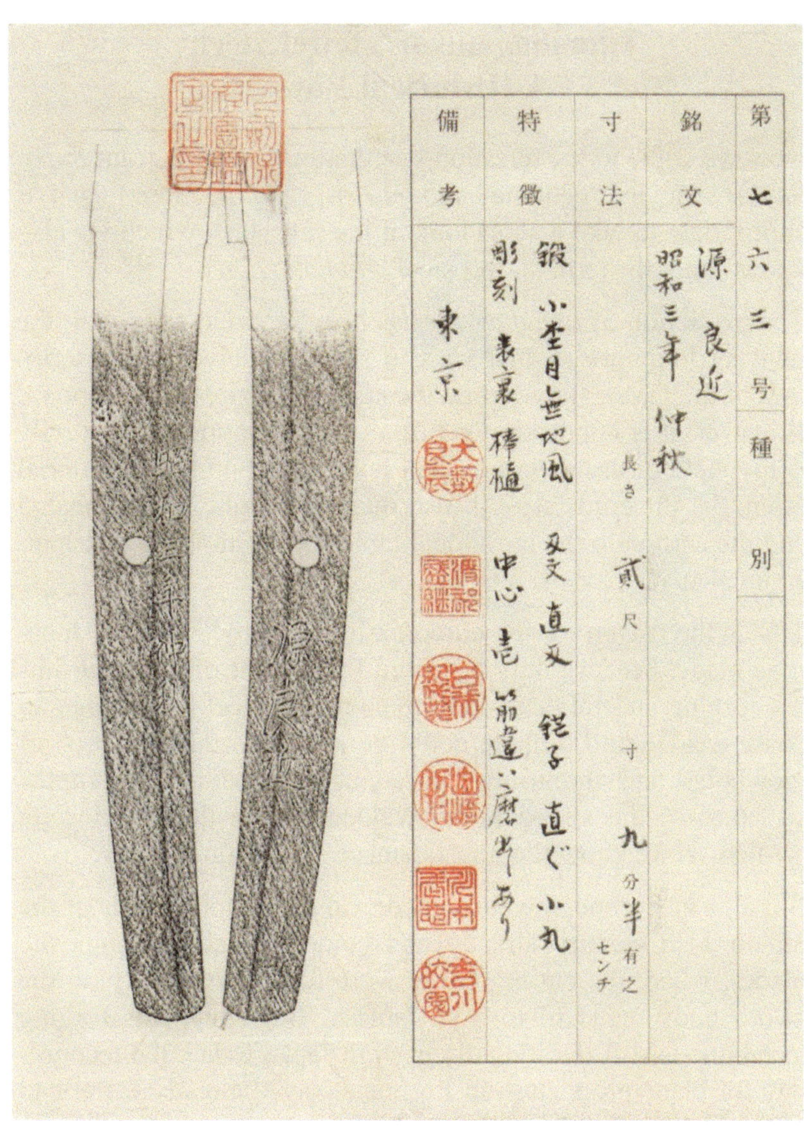

Detail from NTHK Shinteisho with tang abrasion (oshi-gata) and the personal seals of six appraisers.

Tamahagane or „Jewel Steel"
A Historical View

Now we come to the question whether only swords from *tama-hagane* can be "genuine" *nihonto* or *gendaito*. For this it is appropriate to take a short look at the raw steel, which we also know under the term "jewel steel": *tamahagane*.

The extraction of *tamahagane* has been reported enough in the relevant literature so that we can spare ourselves the description of the *tatara* furnace and the smelting process at this point. However, it is important for further consideration that not only is the yield of usable crude steel low in relation to the material input, but the crude steel is also relatively highly contaminated and the carbon in the individual tamahagane lumps is distributed unevenly.[34]

This is the reason for the complex forging process of the Japanese blade, because only by careful treatment of the steel during forging and frequent folding could the steel be homogeneously welded and a blade could be produced that was free of blowholes and impurities. In connection with the complex structure of the sword body, welded composite blades were created, as we know them in Japanese swords until today.

But also in Europe, the early medieval Germanic swords of the 8th to 11th century already had complex welded composite blades, where the cutting edges were forged separately on the sword body made of torsion damask. But while this forging technique is still traditionally used in Japan today, the technology for blade production in Europe changed already at the end of the 11th century.

---

[34] Leon and Hiroko Kapp, Yoshindo Yoshihara, The Craft of the Japanese Sword, Kodansha International Ltd., 1987

Due to the improvement of the smelting furnace technology, the abandonment of complicated manufacturing processes in favor of high-performance mono block blades *(see also Ulf-berht blades)* began at this time.[35] While the *tatara* furnace was traditionally used in Japan, in Europe the smelting technology was permanently advanced and improved.

When Portuguese and Dutch merchant vessels began to increasingly visit the coasts of Japan in the middle of the 16th century, crude steel from Europe for the first time also reached Japan. Japanese smiths quickly recognized the advantages of this steel due to its purity and forged the first swords from *namban-tetsu ("iron from the southern barbarians")*. With the exception of the fact that the raw steel used for sword making was different from *tamahagane*, namely imported steel from Europe, the traditional forging techniques did not change.

Swords made of imported steel were highly coveted by the samurai because of the quality of the steel and fetched higher prices than swords from *tamahagane*. The smiths proudly signed the tangs of these swords with *"namban-tetsu"*. Along with steel came the first rifles *(Tanegashima Rifle)* to Japan, which revolutionized the warfare of that time.[36]

In the historic *Battle of Nagashino* [37] in 1575, the allies under *Oda Nobunaga* and *Tokugawa Ieyasu* first used harquebus shooters in large numbers who took position behind palisades. During the attack, the hitherto undefeated cavalry of the *Takeda* wave after wave was mowed down and almost completely wiped out in the harquebusiers' fire. There are still plenty of swords from this period which, although they were forged

---

[35] http://de.wikipedia.org/wiki/Ulfberht

[36] https://de.wikipedia.org/wiki/Tanegashima-Arkebuse

[37] Goepper, Roger, Oishi Shinzaburo, Tokugawa Yoshinobu, Shogun, Kunstschätze und Lebensstil eines japanischen Fürsten der Sho-gun-Zeit, Katalog zur Ausstellung im Haus der Kunst München, Toppan Printing Co., Ltd., Tokyo, September 1984

from imported steel and not *tamahagane*, have all received *origami* and are therefore recognized as "genuine" *nihonto*.[38]

Under the rule and influence of the mighty Nabeshima Daymio, Hizen province developed from the Keicho period (1596-1615) into a flourishing forging and trading center for new swords made of imported steel. This was favored by the geographical location of Hizen province on the Kyushu peninsula, where there were two prosperous trading ports through which crude steel was also imported from Europe to Japan. One of them was the port of Fukae, today's Nagasaki, which was located directly in Hizen province. Under the monopoly of the Nabeshima Daymio, *Hizen-to*, which was of excellent quality in terms of both artistic and practical qualities, quickly became a sought-after commodity throughout Japan.

Most of these swords come from the Tadayoshi School, which bore many important swordsmiths in the course of its work. The founder of the school was Hashimoto Shinsaemon Tadayoshi. "He went to Kyoto in 1596, where he studied under Umetada Myoju. He returned to Hizen after a three year apprenticeship and lived in the castle town of Saga, where he was employed by the Nabeshima clan. He was later awarded the title Musashi Daijo and changed his name to Tadahiro. Omi Daijo Tadahiro was the second generation. The third, Mutsu Daijo Tadahiro, was later awarded the title Mutsu no Kami."[39]

„In the early years of the first generation Tadayoshi, the jihada is slightly coarse and loose ko-mokume hada mixed with o-hada, but it becomes dense and fine ko-mokume hada with ji-nie and ckikei from the end of the Keicho era. This later jihada

---

[38] "NAMBAN TETSU Project", Token Sugita Europe,
http://www.tokensugita.com/NT.htm
[39] Kokan Nagayama, The Connoisseurs Book of Japanese Swords
Kodansha International, Japan, 1. Edition 1997, p. 246

is called konuka-hada and is characteristic for Hizen-to."[40] The reason for this may have been the use of imported crude steel *(namban-tetsu)*, which, while maintaining traditional forging techniques, opened up completely new possibilities for sword-smiths due to its homogeneity compared to tamahagane. Until the Meiji Restoration, nine generations of Tadayoshi had forged, with up to sixty blacksmiths reportedly working for the second generation alone to meet the great demand for Hizen-to.[41] However, the majority of these swords were still of excellent quality.

The swords of the Edo Echizen Yasutsugu School were just as much in demand. Shodai Yasutsugu was one of the first proponents of imported steel *(namban-tetsu)* in sword making, having recognized its advantages over *tamahagane* early on. Full of pride, he signed the tangs of his swords with the addition *"namban-tetsu"* to indicate the use of imported steel. It is certain that this very fact helped to consolidate and increase his fame as a swordsmith throughout Japan. Many blades of the first generation of Yasutsugu feature elaborately executed *horimono*. It is believed that Yasutsugu personally engraved the *horimono* on his early blades. However, the best known and most beautiful *horimono* on his blades are said to have been made by the famous engravers of the Kinai family.[42]

„Yasutsugu was a native of Omi province and used Shimosaka as his name as a swordsmith in his early years, before moving to Echizen province. When he was given the character "Yasu" from Tokugawa Ieyasu's name, he changed his name to Yasutsugu, and was allowed to use the Aoi-Mon, the family crest of the Tokugawa clan, on his blades. This honor was granted in

---

[40] Kokan Nagayama, The Connoisseurs Book of Japanese Swords
Kodansha International, Japan, 1. Edition 1997, p. 247
[41] Clive Sinclaire, To-Ken Society of Great Britain, "Hizen-to",
http://www.japaneseswordindex.com/hizen-to.htm
[42] https://www.nihonto.com/the-yasutsugu-school

1507 by Tokugawa Ieyasu himself. Yasutsugu alternatively produced swords in Edo and Echizen provinces after being employed by Tokugawa Ieyasu. A struggle over family succession, following the death of the second Yasutsugu, caused a split within the school. As a result, the eldest son came to head the Edo Yasutsugu line of the family, and the third son succeeded to the Echizen Yasutsugu branch. Both used Yasutsugu as their names as swordsmiths for generation after generation, but no smiths equal in skill to the first generation Yasutsugu appeared at any time in the Edo period."[43]

Two blades of the first and third generation Yasutsugu shall finally prove the originality and quality of swords forged from imported steel in ancient Japan. The photos are from the *Japanese Sword Online Museum, Aoi-Art, Tokyo, Japan*, and reflect only a small selection from the Japanese Sword Online Museum's collection.

---

[43] Kokan Nagayama, The Connoisseurs Book of Japanese Swords Kodansha International, Japan, 1. Edition 1997, p. 240

An absolute masterpiece of the first generation Yasutsugu, forged from imported steel, signed "Motte Namban-tetsu Oite Bushu Edo Echizen Yasutsugu" with Tokugawa-mon and Toku-betsu Hozon Origami of the NBTHK.[44] *Photos/oshigata courtesy of Kazushige Tsuruta-San, Aoi-Art, Tokyo, Japan.*

---

[44] https://www.aoijapan.net/katana-motte-nanbantetsu-oite-bushu-edo-echizen-yasutsugufirst-generation/

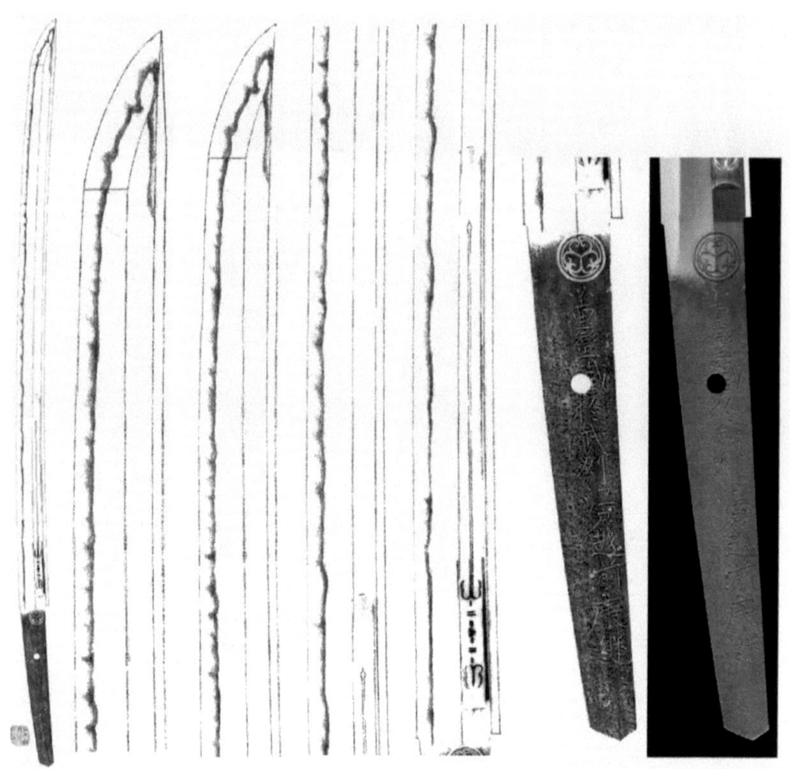

Oshigata of the blade and the tang with photo of the tang of the blade of the first generation Yasutsugu. The coat of arms (mon) of the Tokugawa shoguns can be easily recognized. Shimosaka was personally bestowed with it in 1507 by Tokugawa Ieyasu in connection with the inscription "Yasu". Shimosaka changed his name to Yasutsugu and henceforth signed his swords with Yasutsugu and the Tokugawa-mon. Photos/oshigata courtesy of Kazushige Tsuruta-San, Aoi-Art, Tokyo, Japan.

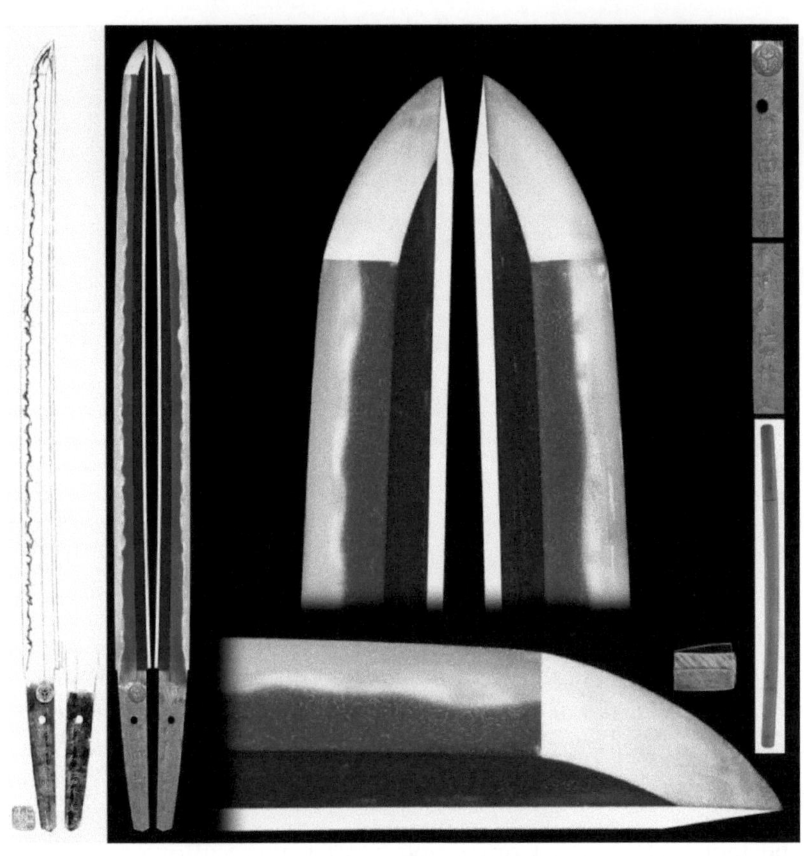

A technically and artistically perfect o-wakizashi of the third generation Yasutsugu made of imported steel in Kanbun-Shinto style, signed "Yasutsugu Motte Namban-tetsu – Oite Bushu Edo Saki No" and Tokugawa-mon. The blade with Tokubetsu Hozon Origami of the NBTHK. [45] Photos/oshigata courtesy of Kazushige Tsuruta-San, Aoi-Art, Tokyo, Japan.

---

[45] https://www.aoijapan.net/wakizashi-yasutsugu-motte-nanbantetsuoite-busyu-edo-saki-no/

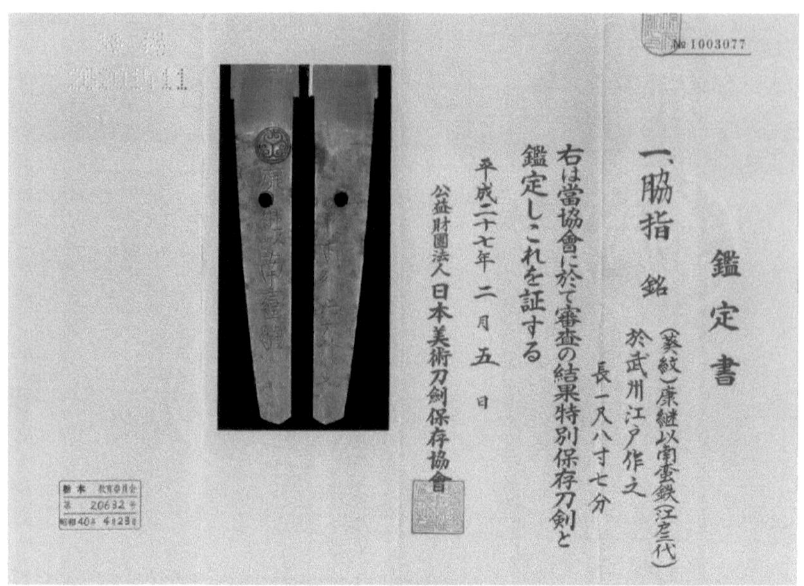

NBTHK Tokubetsu Hozon Origami of the pictured o-wakizashi of the third generation Yasutsugu, signed "Yasutsugu Motte Namban-tetsu – Oite Bushu Edo Saki No" and Tokugawa-mon.[46] Photos/oshigata courtesy of Kazushige Tsuruta-San, Aoi-Art, Tokyo, Japan.

The fact that blades from *namban-tetsu* were allowed to bear the family crest of the Tokugawa shoguns proves the high esteem in which the imported steel was held in ancient Japan. Thus the question whether swords made of imported steel can be "genuine" *nihonto* should be exhaustively answered in favor of imported steel from a historical point of view. This view was already shared by the warriors of feudal Japan, whose lives could depend on a good sword every hour of the day. Full of pride the samurai already wore swords made of imported steel more than 400 years ago, which at that time according to to-

---

[46] https://www.aoijapan.net/wakizashi-yasutsugu-motte-nanbantetsuoite-busyu-edo-saki-no/

day's usage one might have called *"high performance or high-tech nihonto"*.

Minamoto Yoshichika did nothing else when he followed his own path in the further development of Japanese swords towards the *"high performance gendaito"*. His goal was to provide the samurai of the modern era and the officers of His Majesty the Emperor with artistically perfect swords that could withstand the toughest hand-to-hand combat due to their technical perfection. The fact that this in practice was not always true for Japanese swords is a fact that cannot be denied in spite of all appreciation of the Japanese sword.

What follows is not intended to diminish the importance of the Japanese sword as a weapon and a unique work of art in the world, or to deny the Japanese sword the status it holds among all the blank weapons of the world. The following chapter would never have been written if the author himself were not a great admirer of the Japanese sword. Nevertheless, on further reflection we cannot avoid mentioning things that will tear the one or other enthusiast out of his dreams and confront him with reality.

The Symbol of the Broken Sword

"The Japanese sword does not bend or break, and the sharpness of its edge is legendary." Everyone who deals with Japanese swords has heard or read this sentence in this or a similar way. Even though the Japanese sword may be much more resilient than many other blades, we cannot avoid relativizing the belief in the indestructibility of the Japanese sword at this point.

Of course, Japanese swords can also bend. They can also break and although their blades are razor sharp, they are still vulnerable. One reason why the Samurai parried a sword stroke with the back of the blade was to protect the sensitive cutting edge. There are still enough old swords where the back of the blade is literally sawed away from battle scars *(kiri-komi)*. In the same way, the swords were polished differently (sharper) in peacetime than for war, because in peacetime the blade only had to cut through a kimono in a duel. On an opponent in great armor *(oyoroi)* the same edge could easily have been damaged.

The image of the broken sword, which stands for the broken soul of the Samurai, is not by chance. Also in ancient Japan, swords broke in battle if they had not been forged carefully enough and contained forging faults. Likewise, swords can be damaged or destroyed due to improper handling during cutting tests. On the website of the *"Aikido Center of Los Angeles"* there is a remarkable treatise by *Kensho Furuya Sensei* under the title *"Proper Use of Swords"*.[47] *Kensho Furuya Sensei* refers to a rare book by a sword expert who repaired swords for the Japanese army during World War II which had been damaged on the battlefield.

---

[47] Aikido Center of Los Angeles, The Aiki Dojo Newsletter, November 2020, Proper Use of Swords by Rev. Kensho Furuya
http://www.aikidocenterla.com/newsletter

It reports about swords, which were irreparably destroyed. [48]
"Most swords could not withstand the rigors of actual combat and broke. In addition to blades breaking upon impact, he *(the author)* was also critical of the length of the *tsuka* or handle. If the *tsuka* was too long, it easily broke upon impact. In addition, a longer handle would cause the blade to break very often at the "retaining peg hole" *(mekugi-ana)* on the tang. Broken *tsuka* were the second most frequently occurring damages to swords."

It is noted that much of the damage could also be attributed to improper use. "Many soldiers were not properly trained in sword fighting." Furthermore, the causes of damage are seen in the fact that most swords used during the war were less in accordance with the requirements of an ideal fighting sword than with the army regulations. "As a result, many swords were too heavy or were in a disproportion between the length of the tang and the blade, weight, curvature etc."

As an example for an ideal fighting sword *Kensho Furuya Sensei* refers to the swords of the *Muromachi period (1336 - 1573)*. The end of this period was the *Sengoku period (1477 - 1573)*, which went down in the history of Japan as the *"Time of the Fighting Empires"* or *"Time of the Warring Countries"*. At that time the armies no longer met only on the battlefield. The war covered the whole country and was also carried into the cities, where close combat raged in a confined space. Accordingly, fighting techniques and the length and shape of swords changed. The *uchigatana* with an average length of 60 cm was born. [49]

"During the Muromachi period, the beginnings of Iaido, most Iaido blades were usually only about 61 cm long. This also corresponds to the teachings I *(Kensho Furuya Sensei)* have

---

[48] See also pp. 89 ff, Report of Hikosaburo Kurihara
[49] https://de.wikipedia.org/wiki/Uchigatana

just mentioned. The handle of a 61 cm blade is only about 20 cm long – very short for modern standards – but if we examine the early *koshirae* of this period, we find that the sword handle *(tsuka)* is generally about this length. This is not because the early Japanese had smaller hands, but because the length of the hilt is determined by its most effective ratio to the length of the blade".

Kensho Furuya Sensei further reports about one of his Iaido teachers, who had worked as an inspector for the Imperial Japanese Army and who had fought several hand-to-hand combat for life and death with the sword during the war. He was surprised when he learned that his teacher had used a very light, short sword instead of a *katana*, the blade of which was only 23 inches (approx. 58.4 cm) long. It did not have a long, wide blade as is often used by martial artists today.

His teacher explained to him that in exercises it was justifiable to use a long, heavier blade to develop his technique, but in a fight to the death a shorter, lighter blade would be faster and better to use. A lighter blade would cut just as well and would not break as easily as a long blade. In the same way, *Hakudo Nakayama* is said to have recommended shorter blades for man-to-man combat.[50]

---

[50] Aikido Center of Los Angeles, The Aiki Dojo Newsletter, November 2020, Proper Use of Swords by Rev. Kensho Furuya
http://www.aikidocenterla.com/newsletter

The Japanese Sword as an Art Object

Before we continue with the reflection of the swords of Mina-moto Yoshichika, it is necessary to explain briefly, what are the essential characteristics that distinguish the Japanese art sword from the pure weapon. To this end, what Prof. Otto Kümmel, founder and director of the Museum of East Asian Art in Berlin, who traveled to Japan from 1906 to 1909, wrote about this in his book "Das Kunstgewerbe in Japan"[51] is still valid:

"The essential element (of the Japanese sword) is of course the blade... Meanwhile, the characteristics of the Japanese blade are so deeply hidden under the surface... that it is almost impossible for a European to come to an understanding of even coarser differences. All Japanese blades that rise above a certain, very low level seem at first to the European almost incomprehensibly perfect, and the clear quality levels that every Japanese with some sword education sees at first glance are (for the European) almost unrecognizable."

Michael Hagenbusch explains this on the occasion of the first European symposium on the topic "The Art of the Samurai"[52]: "First of all, one thing is important here, namely to make it clear what these differences in quality are all about. Perhaps the translation of the name of the Japanese society, the already mentioned NBTHK, may contribute to this. Translated this society is not only called "Society for the Preservation of the Japanese Sword", but "for the Preservation of the Japanese Art Sword". In order to fulfill this characteristic, a sword must "possess both artistic and historical value, have its own unique and outstanding features in terms of form, quality and technical

---

[51] Kümmel, Otto, Das Kunstgewerbe in Japan, Schmidt & Co. Berlin, 3. Auflage 1922

[52] Hagenbusch, Michael, Katalog zum ersten europäischen Symposium „Die Kunst der Samurai", Deutsches Klingen-Museum, Solingen 1984

details, and be representative of a particular artist, school or place where it was forged."

"These characteristics of the Japanese sword can be clearly recognized and learned without further ado, since they are obvious – trained eyes presupposed. ...The same way of recognizing and classifying architecture, painting, sculpture or music by its style is also applicable to the sword, which can be recognized by its style and classified in time and provenience, whereby one has more criteria at one's disposal than in other arts, since the sword combines several art forms." Michael Hagenbusch continues: "Japanese blades are high art, an art which the viewer will take a certain pleasure in studying. Furthermore it is necessary to memorize the characteristics of the different schools, provinces and masters, one must be able to say to a blade: I know you, I know where, in what time and by whom you were made in order to really come to the understanding of the certain differences in quality."

The uncertainty of many collectors to be able to say to a blade "I know you, I know when, where and by whom you were made" is often expressed in the request for *origami* when buying a sword. "Does it have papers?" is the most frequently asked question, even before the prospective buyer has dealt with the quality and aesthetics of the sword. Precisely because collectors with a lack of sword education cannot say when, where and by which smith a blade was forged, they demand a certificate that says what they themselves do not know.

With restriction the demand for papers becomes comprehensible, if it concerns a sword, which is not signed *(mumei)* or lost its signature in the course of its history by shortening the tang *(suriage nakago)*; likewise with the acquisition of very expensive swords, with which the costs for certification are negligible in relation to the purchase price or if the suspicion of a wrong signature *(gimei)* arose.

This is different with *gendaito*, where the costs for presenting the sword for inspection in a *shinsa* and obtaining the desired certificate including transport costs, insurance, customs fees and fees for issuing the police registration – in Japan swords are treated similarly to firearms and require police approval and registration – can often amount to another quarter or a third of the market value of the blade offered for sale.

At *gendaito* there are enough possibilities to inform oneself about the importance of a swordsmith and his work with the help of the relevant literature and serious publications on the internet and to study the style of the master in the execution of his signature with the help of *oshigata*. The best way to do this, of course, is to take the desired blade in your hand and study it extensively in order to form your own opinion about the style and artistic execution of the blade as well as the aesthetics of the tang and the authenticity of the signature.

Art Swords for the Hardest Close Combat

When you pick up a sword from Minamoto Yoshichika, you immediately notice the amazing balance. One has the impression that the sword begins to live in the hand and wants to move. Minamoto Yoshichika achieves this phenomenon through the slim, balanced shape of his blades and the fact that he additionally provides them with perfectly cut fullers *(bo-hi)* on both sides. This profile makes the blades lighter while at the same time increasing their torsional strength. Combined with the deep curvature *(sori)*, the high central ridge line *(shinogi)* and the functional tempering line *(hamon)*, Minamoto Yoshichika thus created blades whose cutting performance is legendary. All these technical details and the excellent steel quality made his swords formidable weapons.

By the way, cutting *bo-hi* is associated with a risk that not every swordsmith wanted to or was able to take. In this metal-cutting process, the swordsmith uses a drawing knife to cut into deeper layers of steel. Only a swordsmith who is absolutely sure that he has welded all layers homogeneously can cut deep into the steel without causing cracks and making forging defects visible. This is why *bo-hi* cut to perfection on traditionally forged blades is a testament to the mastery and skill of the swordsmith. Conversely, engraving was and is sometimes used to conceal superficial flaws (break-outs) on blades.

In addition to the technical details that make Minamoto Yoshichika's swords perfect weapons, his work in the execution of *sugata*, *hada* and *hamon* and the design of the tang, which is always another important indicator of the quality of a blade in Japanese swords, prove his high level of craftsmanship and artistic ability and in his work he revives the golden age of the Japanese sword. His swords are unmistakable in form, technical details, quality and artistic execution and represent his own style and outstanding mastery. Due to his appointment as an Imperial Swordsmith and his works on the occasion of the coronation ceremonies of Emperor Hirohito as well as his close correlation to *Hakudo Nakayama*, Minamoto Yoshichika and his works undoubtedly also have a special historical significance.

When examining his blades, it becomes clear why Minamoto Yoshichika is one of the most important sword smiths of his time. In which we take a sword of Minamoto Yoshichika in the hand, in the truest sense of the word we "grasp" that his work not only meets the high aesthetic demands on the Japanese sword as an object of art, but that by producing fast, highly durable and extremely sharp blades, he has also perfectly realized the requirements of experienced swordsmen for an efficient weapon. Thus, Minamoto Yoshichika has ideally combined art and usability in his works and thus created swords of extraordinary perfection and high value. John Scott Slough and Tokuno Kazuo summed up this achievement when they classified Minamoto Yoshichika's swords as *"High Grade Gendaito"*[53] and *"Highest Grade Gendaito"*[54].

Minamoto Yoshichika, Shodai and Nidai, are among the most important swordsmiths of the Taisho and Showa period, who created swords with art status for Japan's "last samurai" on their way into the material battles of the 20th century, which were able to withstand even the most extreme stresses in the hardest close combat. Their swords thus clearly stand out from the mass of swords forged during this period. Considering the fact that the two generations of Minamoto Yoshichika forged relatively few swords and that blades of the Nidai are rare anyway, it is understandable that swords with the signature *"Minamoto Yoshichika"*, no matter if Shodai or Nidai, are now intensively sought after and highly coveted by collectors. Last

---

[53] Slough, John Scott, An Oshigata Book of Modern Japanese Swordsmiths 1868 – 1945, Rivanna River Company, 2001
[54] Tokuno, Kazuo, TOKO TAIKAN, YOS1067, 2004

but not least are also swords, where we find a cutting test on the tang, rather rare and generally something special. This applies to antique swords and even more so to gendaito.

Minamoto Yoshichika, Shodai and Nidai
Picture Section

While the gendaito of the Taisho and early Showa period were long ignored by conservative collectors, intensive research into the history of these swords led to a rethink and the works of important smiths of the Taisho and early Showa period became sought-after collector's items. Naturally, the focus was primarily on the blades. In the beginning much less attention was paid to the mounts. Worse still, while the blades were often expensively repolished and preserved in newly manufactured *shirasaya*, the original mounts were often disposed of as worthless accessories.

Today we know that good shin-gunto or kai-gunto mountings are handcrafted with great effort and have their own monetary value.[55] But what makes these mounts special and thus absolutely worth preserving is their cultural-historical significance. A sword reveals its original purpose only with the corresponding military mount. The mount also tells us where the former wearer served in the armed forces, allows us to draw conclusions about his military rank, sometimes also about his origin, encourages us to further research and thus brings contemporary history to life. The following swords are all in their original mountings.

---

[55] http://www.jp-sword.com/files/gunto/gunto-sale.html

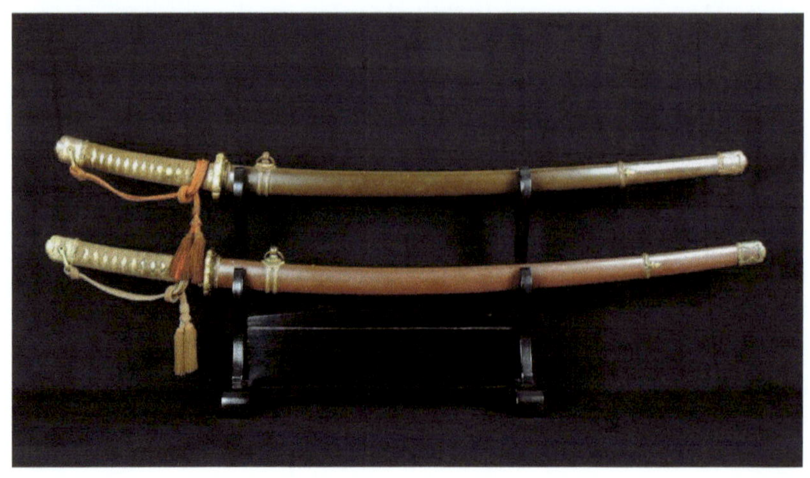

Minamoto Yoshichika, Shodai and Nidai. The Shodai in Gunto mounting type 94 Shin-Gunto, introduced in 1934, below the Nidai in Gunto mounting type 98 Shin-Gunto, introduced in 1938.

Two swords of Nidai Minamoto Yoshichika; above in Gunto mounting type 98 Shin-Gunto, below in Gunto mounting type Navy Tachi Gunto, introduced 1937.

Tsuka of the gunto mountings type 94 and 98 Shin-Gunto. The different material thickness of the tsuba is clearly visible. In both cases, the tsuba is still a detailed sukashi tsuba (openwork tsuba). The mounting type 94 Shin-Gunto corresponded to the classical tachi mounting and had two support rings, whereby the front support ring was fixed. The rear support ring was removable, as the sword was only worn on the front ring during service, and is therefore often no longer present. The Type 98 Shin-Gunto mounting was equipped with only one support ring from the beginning due to the way it was carried.

Shodai Minamoto Yoshichika
with Cutting Test by Hakudo Nakayama

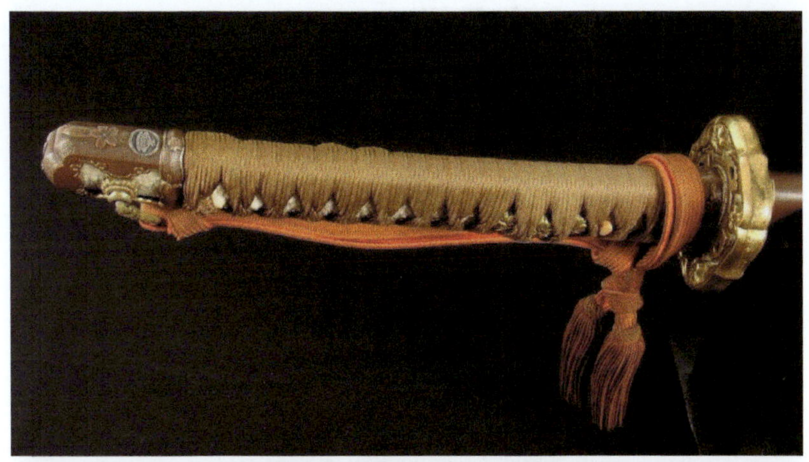

The heavy, gold-plated tsuba type 94 Shin-Gunto and the kabuto-gane in nanako technique with cherry blossom decoration speak for the quality of the mounting. On the kabuto-gane the silver family crest of the sword bearer in the form of a command fan. An indication that the officer came in a direct line from a samurai family.

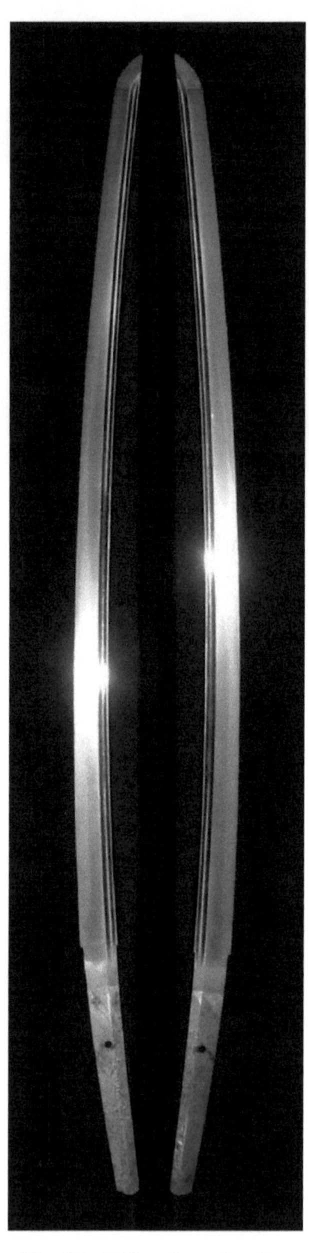

A Shodai Minamoto Yoshichika for the Imperial Guard. A rare sword with cutting test by Hakudo Nakayama.

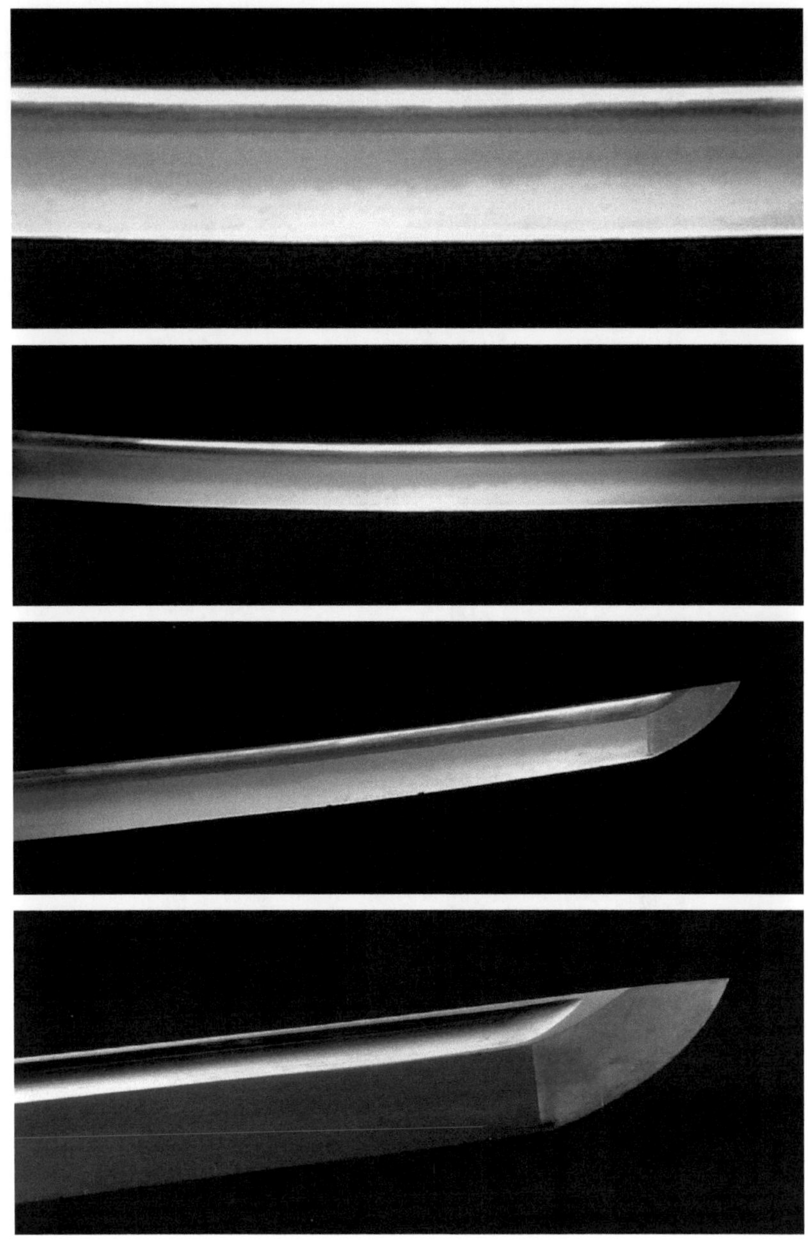

The tang (nakago) katana-mei with caligraphically beautiful executed signature "Minamoto Yoshichika", dated "Showa Ju Nen Ni-Gatsu-Hi" ("A day in February 1936"); below the cutting test of Hakudo Nakayama.

A work by Shodai Minamoto Yoshichika with cutting test by Hakudo Nakayama. The blade shinogi tsukuri with iori-mune, ko-kissaki and perfectly cut bohi. Hada very dense ko-mokume, hamon midareba with notare, gunome midare and kochoji in nioi deki, boshi midare komi. Nagasa 669 mm, sori 13 mm, motohaba 30.0 mm, sakihaba 20.0 mm, motokasane 6.5 mm, sakikasane 4 mm, nakago 188 mm. The stylish taper of the blade width (haba) and blade thickness (kasane) towards the blade tip (kissaki) reminds us of the tachi of the early to middle Kamakura Period.

The blade still with ricasso (ubu-ha) and in first original polish, this slightly gray or "cloudy". Nevertheless, the blade reveals the special quality and beauty that can be expected from a work of Minamoto Yoshichika. In the area of the monouchi, there are two tiny nicks in the cutting edge that could easily be polished out in the course of a new polish. Although the sword is absolutely worth polishing, a new polish was waived to preserve the blade in its authenticity.

The carefully crafted tang with kurijiri, kaku-mune and beautifully applied kesho yasurime. Katana-mei with very beautifully executed caligraphic signature "Minamoto Yoshichika". Ura dated "Showa Ju Nen Ni-Gatsu-Hi" ("Showa tenth year (1936) one day in February") with cutting test by Hakudo Nakayama "Hakudo Tameshigiri Sho" ("Cutting test by Hakudo"). One of the few blades still known, forged by Minamoto Yoshichika for the Imperial Guard and tested by Hakudo Nakayama. The present sword is rare even beyond that and special because it is also dated. Blades by Yoshichika often have no dating on the tang.

Koshirae Type 94 Shin-Gunto[56], official designation "Army commissioned officers Shin-Gunto 1934". Outer saya steel sheet in brown lacquer*. Painting of the saya and all fittings in good to complete condition. Early, 10 mm thick and 172 g heavy sukashi-tsuba with approx. 80% original gilding. Ohmura Tomoyuki notes on his recommended homepage "Ohmura's Gunto Site": "Sukashi tsuba (early version, General's) Brass cast. Bigger and flamboyant".[57] Kabuto-gane with silver family crest (mon)** showing a command fan. Tsuba and all fittings including the seppa probably signed by hand with a drawing pin with the same kanji. Tsuka with excellent same (ray skin), tsuka-ito silk, handle winding morohineri-maki style, gold plated copper habaki. Overall very good quality of the koshirae. Red-brown sword knot (tocho)*** added later.

*The saya of Type 94 Shin-Gunto consisted of the outer sheath made of steel or aluminum sheet and a wooden liner made of magnolia wood in which the blade ran. The paint was brown.[58]

**An officer was allowed to place a family crest (mon) on the sword hilt (tsuka). The attachment was mainly on the kabuto-gane or a menuki and was an indication that the officer came in a direct line from a samurai family.

***Blue-brown sword knots with a blue-brown tassel were worn from the lieutenant to the captain ("company grade"), red-brown sword knots with a red-brown tassel from the major to the colonel ("field grade") and red-brown, zig-zag gold-twisted sword knots with golden tassels were worn by generals ("generals grade"). Marine officers wore uniform brown sword knots with brown tassels.[59]

---

[56] http://ohmura-study.net/931.html
[57] http://www.jp-sword.com/files/gunto/parts.html
[58] http://ohmura-study.net/909.html
[59] http://www.jp-sword.com/files/gunto/tassel.html

The cutting test by Hakudo Nakayama and the family crest on the kabuto-gane prove that the bearer of the sword came in a direct line from a samurai family and served as an officer in the Imperial Guard.[60] The fact that Minamoto Yoshichika, Imperial Swordsmith and the only swordsmith of the Taisho period who found mention into Fujishiro's "Nihon Toko Jiten, Shinto-hen" created this blade and Hakudo Nakayma, Budo Grand Master, Meijin and the most famous swordsman of his time also held this blade in his hands to subject it to a cutting test seven times before he accepted it for the Imperial Guard, makes this sword not only important in terms of cultural history. With a deeper study, the blade reveals to the observer the genius and brilliance of two men who were among the most important sword greats of their time.

---

[60] Fuller, Richard and Gregory, Ron, Military Swords of Japan 1868 – 1945, Arms and Armour Press, London - New York - Sydney, 1986

Nidai Minamoto Yoshichika
in Gunto Koshirae

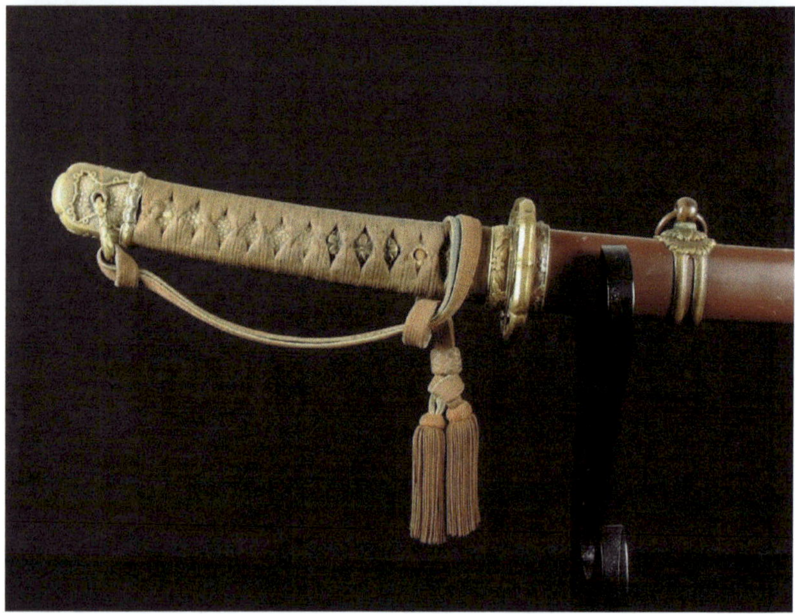

The worn mounting in overall good condition.

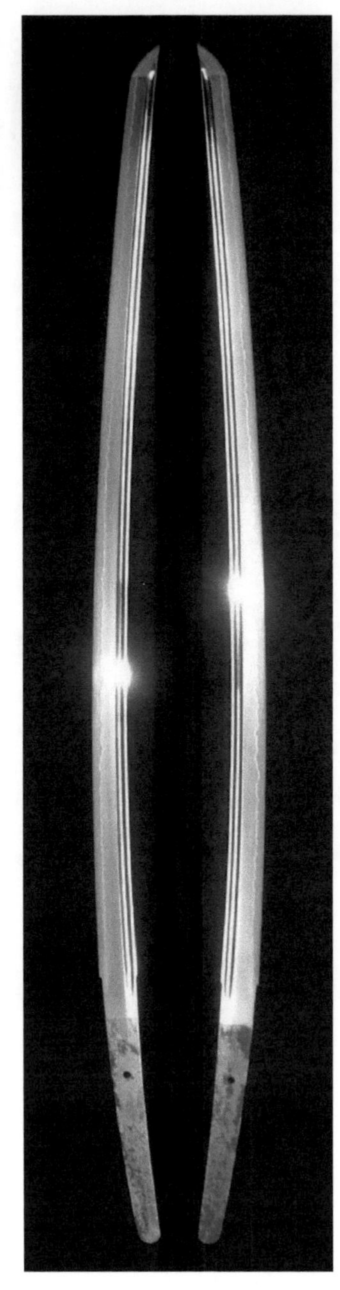

67

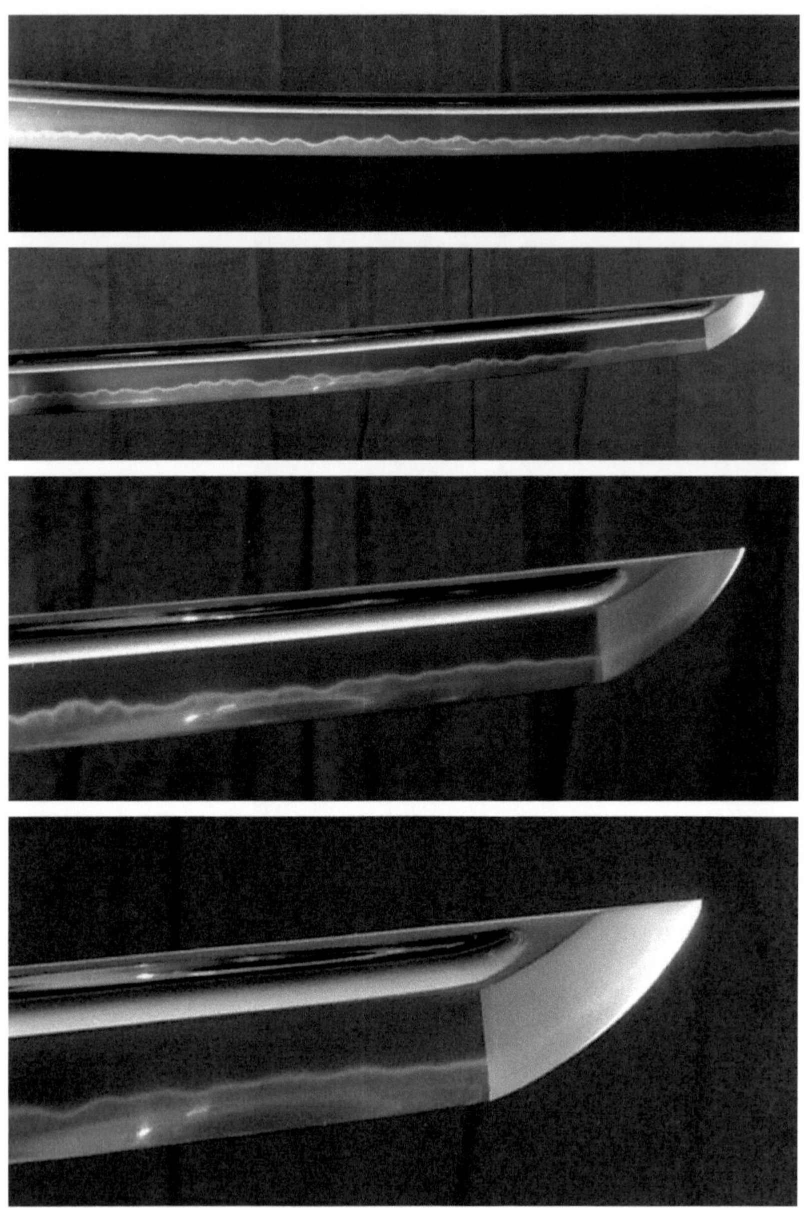

*Dark shadows near the cutting edge are reflections on the ab-
solutely perfect and flawless blade.*

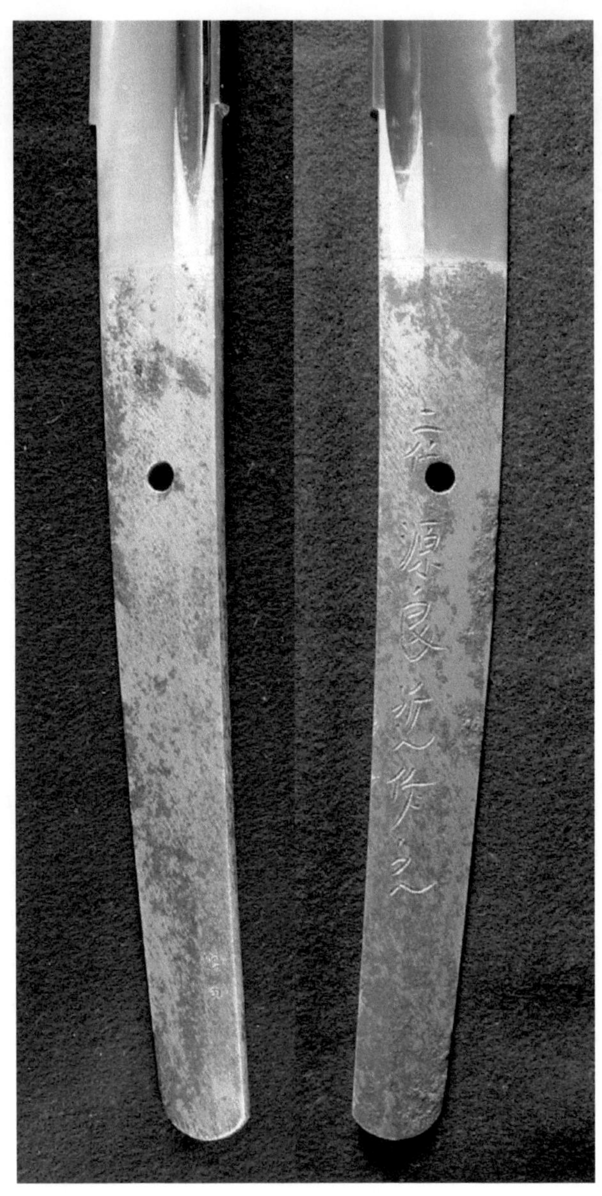

Nakago signed tachi-mei "Nidai Minamoto Yoshichika saku kore". On the ura a tiny logo of the "Suya Sho Ten" and a small Arabic "6".

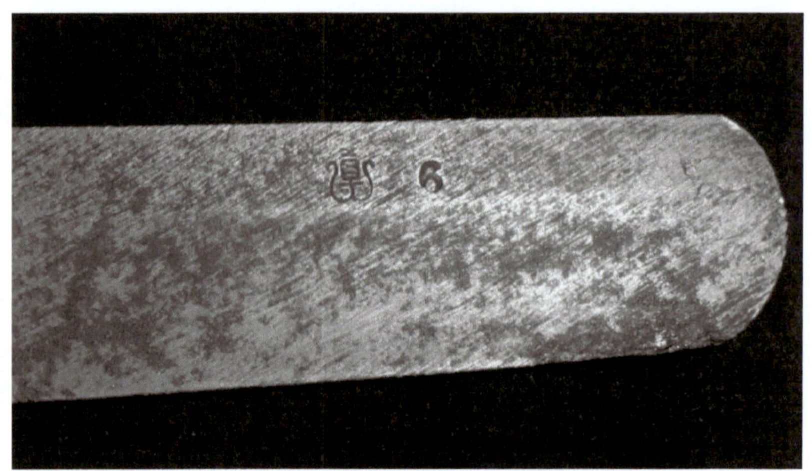

The logo of the "Suya Sho Ten and the Arabic "6".

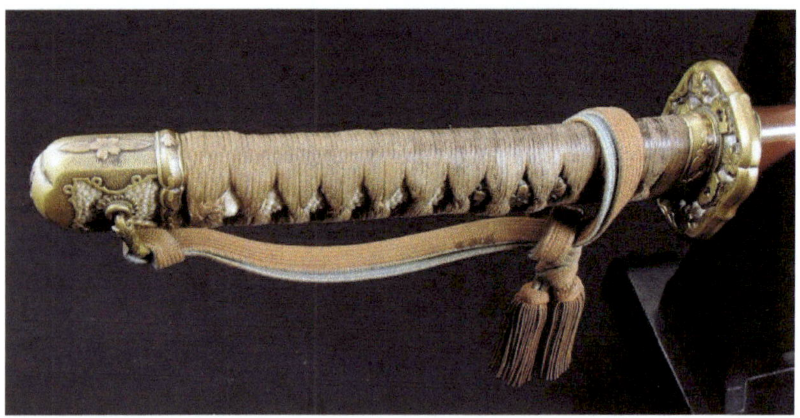

The handle (tsuka) of the mounting type 98 Shin-Gunto with beautiful openwork tsuba. Kabuto-gane in nanako technique and cherry blossom decoration. The mount was made by the "Suya Sho Ten", which traditionally worked for the imperial court, high-ranking officers and diplomats.

An artistically outstanding work of Nidai Minamoto Yoshi-chika in new polish. The blade shinogi tsukuri with iori-mune, ko-kissaki and perfectly cut bohi convinces by its elegant form and the outstanding quality of hada and hamon. Hada very dense masame, hamon sugu komidare with gunome and kocho-ji in nioi deki with countless ashi and yo, boshi suguha komaru kaeri. The blade shows utsuri and proves once more the special mastery of the Nidai.

The carefully crafted tang with kurijiri, kaku-mune and beauti-fully applied kesho yasurime, signed tachi-mei "Nidai Mina-moto Yoshichika saku kore" ("Nidai Minamoto Yoshichika did this"). On the ura a tiny logo of the "Suya Sho Ten" and a small Arabic "6". Nagasa 675 mm, sori 13 mm, motohaba 29,5 mm, sakihaba 20,0 mm, motokasane 6,2 mm, sakikasane 5 mm, nakago 195 mm.

Koshirae Type 98 Shin-Gunto, official designation "Army commissioned officers Shin-Gunto 1938".[61] Saya in brown lacquer* and made of aluminum sheet, detailed sukashi-tsuba. Tsuba and all fittings including all seppa probably signed by hand with a drawing pin with the same kanji. Tsuka with excel-lent same (ray skin), tsuka-ito silk, handle winding morohineri-maki style. Army style gold plated copper habaki. Overall very good quality koshirae in good condition. Blue-brown sword knot (tocho)** added later.

*The saya of Type 98 Shin-Gunto consisted of an outer sheath made of steel or aluminum sheet and a wooden liner made of magnolia wood in which the blade ran. The standard paint was green. Custom ordered mounts had a brown finish.[62]

**Blue-brown sword knots with a blue-brown tassel were worn from the lieutenant to the captain ("company grade"), red-brown sword knots with a red-brown tassel from the major to

---

[61] http://ohmura-study.net/934.html
[62] http://ohmura-study.net/909.html

the colonel ("field grade") and red-brown, zig-zag gold-twisted sword knots with golden tassels were worn by generals ("generals grade"). Marine officers wore uniform brown sword knots with brown tassels.[63]

1999, the then owner had asked the honorable *Mr. Han Bin Siong (* 21.08.1932, † 09.03.2005)* for his estimated judgement to the sword. *Han Bing Siong*, author of the renowned documentation *„Japanese Swords in Dutch Collections"*[64], is still regarded in professional circles today as an expert and undisputed authority and connoisseur of the matter due to his profound knowledge. His expressive *oshigata*, which he also prepared for the catalog of the *First European Symposium "The Art of the Samurai"*[65], were and are always a valuable help for beginners and advanced students in the study and classification of Japanese blades.

The former owner was worried that the stamp on the tang was an arsenal stamp and could be an indication that the blade might not have been forged traditionally. In his letter of September 2, 1999, Han Bing Siong dispelled this concern by first of all congratulating the then owner on his "Nidai Minamoto Yoshichika in gunto koshirae". He goes on to say that the stamp had been of unknown origin for a long time and that Jim Dawson[66] had only recently found out that it was the logo of the *"Suya Sho Ten"* (English "suya company"). The company was founded in the *Meji era* by a certain Mr. Shimada and has a long tradition of manufacturing *koshirae*. Among others, the

---

[63] http://www.jp-sword.com/files/gunto/tassel.html
[64] Han, Bing Siong, Japanese Swords in Dutch Collections: A Selection from 500 Descriptions in the Series Japanse Zwaarden in Nederlands Bezit, De Nederlandse Tōken Vereniging, 2003
[65] Katalog zum Ersten Europäischen Symposium „Die Kunst der Samurai", Deutsches Klingen-Museum, Solingen 1984
[66] Dawson, Jim, Swords of Imperial Japan, 1868 – 1945, Stenger-Scott Publishing, 1996

company worked for such prominent clients as the Imperial Family, many high-ranking officers and diplomats, and the Naval Supply Center *(Suikosha).*[67]

Richard Fuller and Ron Gregory, authors of the standard work *"Military Swords of Japan"*[68], have the Suay Sho Ten logo on page 81 under "xvii". In the accompanying note on page 80, the 1986 edition still reads: "Unidentified; appears to be related to the Kokura Arsenal."[69]

Han Bing Siong further writes that he himself owns a blade (probably in *Kyu-Gunto-koshirae*, in the letter "sabre"), which is also stamped with this logo. He further explains that this sword is the first traditionally forged *katana* with such a stamp on the tang that he has heard of.

Everything indicates that the present blade was a commissioned work, forged by Nidai Minamoto Yoshichika and mounted by the Suya Sho Ten. This is indicated by the special quality of the mounting and the lacquering, which was "custom ordered"; as well as the saya made of aluminum sheet, which served to save weight. In the further course of the war it was abandoned due to the increasing scarcity of raw materials, as aluminum sheet was needed for aircraft construction.

Page 74: The letter from Mr. Han Bing Siong to the then owner of the Nidai Minamoto Yoshichika quoted in the text. The name of the addressee has been obscured for privacy reasons.

---

[67] http://www.japaneseswordindex.com/logo/logo.htm
[68] Fuller, Richard and Gregory, Ron, Military Swords of Japan 1868 – 1945, Arms and Armour Press, London - New York - Sydney, 1986
[69] https://guns.fandom.com/wiki/Kokura_Arsenal

September 2, 1999

Dear Mr   ,

Nice to hear from you. Congratulations on having a Mukai Yoshichika in gunto koshirae.

Up until recently the stamp on your sword was of unknown origin, provided it is like this [seal]. Recently Jim Dawson found out that it is the logo of the Suya Company. It is often seen on the tsuba of shin gunto and also of the 1944-gunto. I have a sabre with the blade also stamped with that logo. Your sword is the first traditional style katana I have heard of with such a stamp on the nakago.

Hoping that this information will be of use to you

Best regards

Nidai Minamoto Yoshichika
in Kai-Gunto Koshirae

The tsuba with stylized rising sun is typical of the abstraction and reduction to the essential, as we know it from Japanese heraldry.

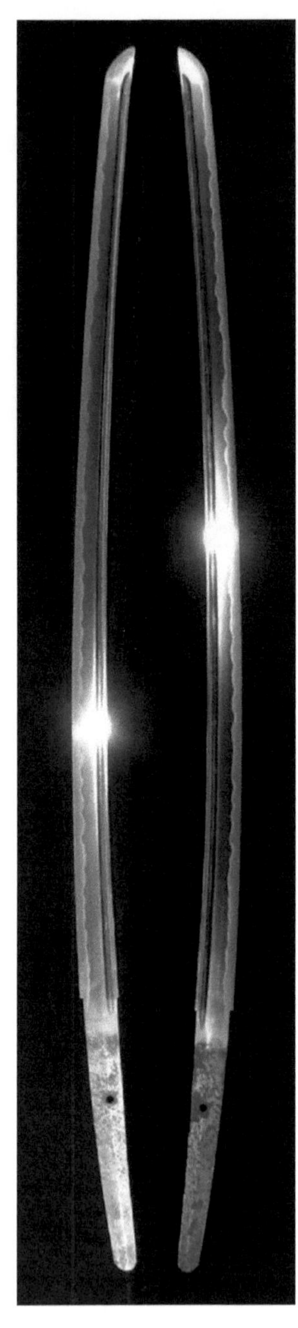

76

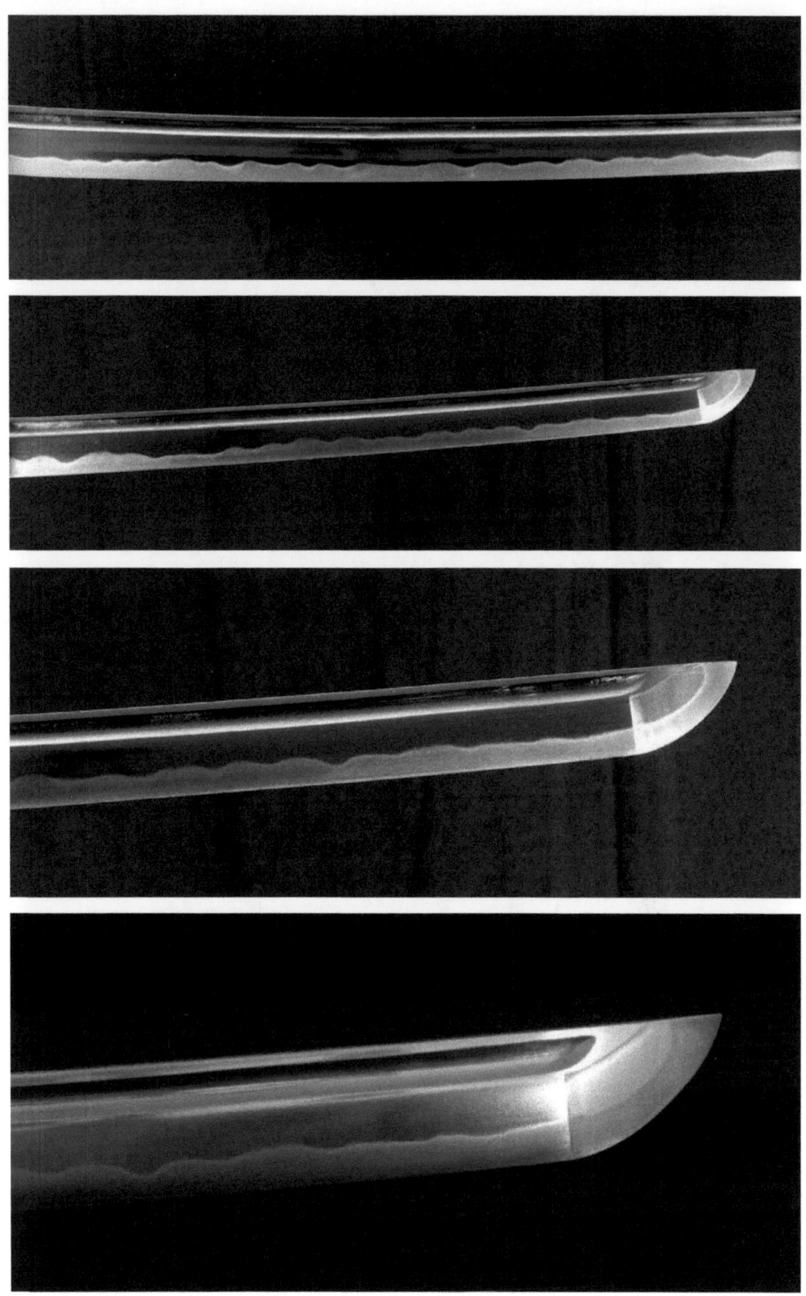

Tang signed tachi-mei "Nidai Minamoto Yoshichika kore saku".

78

Handle (tsuka) with black lacquered ray skin (same) and handle winding in hiramaki style. Brown sword knot (tosho) standardized for all naval officer ranks.

Naval menuki with silver family crest (mon) depicting a crane.

Handle (tsuka) in good, worn condition with nice patina.

The scabbard (saya) made of magnolia wood with leathering of finely textured ray skin (same).

A typical work for Nidai Minamoto Yoshichika in kai-gunto koshirae. The blade shinogi tsukuri with iori-mune, ko-kissaki and bohi convinces by its elegant shape and the quality of hada and hamon. Hada extremely dense ko-itame, almost muji hada. Hamon notare gunome with kochoji in nioi deki, similarly active with many ashi and yo as in the previously described work of the Nidai in shin-gunto koshirae. O-maru boshi.

The carefully crafted tang (nakago) with kurijiri, kaku-mune and beautifully applied kesho yasurime, signed tachi-mei "Nidai Minamoto Yoshichika kore saku" ("Nidai Minamoto Yoshichika did this"). Nagasa 679 mm, sori 13 mm, motohaba 28.5 mm, sakihaba 20.0 mm, motokasane 6.5 mm, sakikasane 5 mm, nakago 195 mm.

Koshirae type "Navy Commissioned Officers Shin-Gunto 1937", common name "Navy Type Tachi Gunto"[70]. Saya made of magnolia wood with black painted ray skin. Beautiful heavy shakudo-plated naval tsuba with stylized rising sun on the o-seppa. All fittings are detailed and of good quality. Tsuka with black lacquered same (ray skin), tsuka-ito silk, handle winding hiramaki style. Menuki on the omote with silver family crest (mon)* depicting a crane. Navy style silver plated copper habaki. Original brown Japanese naval sword knot (tocho). Overall nice Kai Gunto Koshirae in good condition.

*An officer was allowed to place a family crest (mon) on the sword hilt (tsuka). The attachment was mainly on the kabuto-gane or a menuki. The mount and the family crest prove that the bearer of the sword served as an officer in the Imperial Navy and came in a direct line from a samurai family.[71]

---

[70] http://ohmura-study.net/909.html
[71] Fuller, Richard and Gregory, Ron, Military Swords of Japan 1868 – 1945, Arms and Armour Press, London - New York - Sydney, 1986

靖國神社
The Yasukuni Shrine

"On the grounds of the Yasukuni Shrine near Kudan in Tokyo, between a sumo ring and a pond with a fountain, there is a new, dignified building reminiscent of an old samurai residence. This building is the swordsmith's factory (or workshop) of *Nihonto Tanren Kai (Japanese Swordsmithing Center)*. As you approach the pond, a subtle rhythm is in the air – the sound of forging swordsmiths." This is how *Kurata Shichiro*, manager of the *Nihonto Tanren Kai Foundation*, begins his article about the forging center at Yasukuni Shrine, which appeared in the October 1933 issue of *Token Kai Magazine*.[72] The main task of the foundation, which was established in July 1933 by the Japanese Minister of War *Araki Sadao* with the help of a group of selected swordsmiths, is to supply the officers of the Imperial Army with high-quality samurai swords. Capacity reserves are used to carry out orders for the government, shrines, temples, official organizations or special personalities. "The Forging Center also promotes the livelihood of the swordsmiths and encourages them to develop their skills while cultivating the love of the Japanese sword and promoting the fighting spirit in the heart of the nation." So far *Kurata Shichiro* in his article.

The Yasukuni Shrine is a *Shinto shrine* dedicated to the souls of all the soldiers killed in action who have given their lives on the side of the Imperial Army since the *Meiji Restoration*, but also to the war dead of all nations, including the enemies of war. The shrine was originally erected in 1869 under the name *Tokyo Shokonsha (Tokyo Shrine for Summoning the Spirits of the Dead)* in memory of all war dead and heroic souls.

---

[72] Tom Kishida, The Yasukuni Swords, Rare Weapons of Japan 1933 – 1945, Kodansha Europe Ltd., 1. Edition 2004, p. 50

Japanese marines visiting the Yasukuni Shrine. In the foreground left on the picture an officer with presented samurai sword receives the message.

In 1879 *Emperor Meiji* elevated him to *Bekkaku Kanpeisha (Imperial Shrine of the Special Class)* and gave him the name *Yasukuni-jinja ("Shrine of the peaceful country")*. The shrine is visited every year by the bereaved of the fallen and veterans' associations, but also by nationalist associations. Official visits to the shrine by politicians repeatedly enrage Japan's neighbors and promptly lead to official protests. For the Chinese and Koreans, the Yasukuni Shrine is a memorial to the glorification of the darkest chapters of Japanese history. Particularly harsh criticism is levelled at the fact that the officers sentenced to death at the war crimes trials in Tokyo and members of the infamous Unit 731, which conducted experiments with biological weapons on prisoners of war and Chinese civilians in occupied Manchuria, were also included in the list of the *kami.*

Before they go to war, soldiers of the Imperial Japanese Army visit the Yasukuni Shrine.

Right-wing circles resist this criticism by pointing out that this is not a war memorial in the sense of nationalist propaganda, but a shrine where the angry souls of the deceased are to be appeased so that they do not cause discord in the country. This opinion is also supported by international research on Japanese cultural history.[73] Left-wing critics, on the other hand, call for a rejection of all forms of militarism and a critical approach to their own history in the face of Japanese war crimes.

International criticism is directed primarily against visits to the shrine by high-ranking politicians. In Japan, an attempt is made to defuse the situation by making a subtle distinction between private visits (the fathers of these politicians are venerated as war dead in the shrine) and visits in an official business. How-ever, the external effect remains the same, so that there are

---

[73] http://de.wikipedia.org/wiki/Yasukuni-Schrein

repeated protest reactions from Korea and China.[74] Thus, the Yasukuni Shrine is not only of religious and special cultural importance, but also of explosive political significance to this day. But back to the history of the *Nihonto Tanren Kai* and his swordsmiths.

The swordsmiths who worked at the shrine were called *"Yasukuni-tosho"* and were given a name when they were appointed as shrine smith, preceded by the first two syllables of the shrine name: *Yasutoku, Yasumitsu, Yasutake, Yasunori, Yasutoshi, Yasuoki, Yasunobu, Yasushige, Yasuyoshi, Yasuaki, Yasumune, Yasukuni* and *Yasuhiro*. Shimazaki Yasuoki was one of these *13 Yasukuni-tosho* who worked in *Nihonto Tanren Kai* on the sacred ground of Yasukuni Shrine in Tokyo from 1933 to 1945. Born in 1916 in *Kyochi Prefecture*, Yasuoki first entered the forge center in 1935 as a charcoal crusher *(sumikiri)* and in the same year became Yasutoku's journeyman *(sakite)*. In 1940 he was awarded the title of *"Yasukuni-tosho"* and the name *"Yasuoki"*. Until 1945 he forged about 750 blades at the shrine. After a fulfilled life as a swordsmith, Yasuoki died in 1986 at the age of 70. In *Tom Kishida's* recommendable book about the shrine and his swordsmith *"The Yasukuni Swords, Rare Weapons of Japan 1933 – 1945"* we find impressive photos, which show Yasuoki still at the age of over 65 years concentrated at work.[75] But before discussing one of his special works, we should first mention the tremendous efforts made by the foundation to run the forging center at Yasukuni Shrine.

In order to produce the large quantity of swords that the Imperial Army needed for its officers, the supply of raw materials first had to be secured. For this purpose, in February 1933, the manager *Kurata Shichiro* started a survey of still existing *tata-*

---

[74] http://de.wikipedia.org/wiki/Yasukuni-Schrein
[75] Tom Kishida, The Yasukuni Swords, Rare Weapons of Japan 1933 – 1945, Kodansha Europe Ltd., 1. Edition 2004, pp. 28 und 29

ra production facilities and the stock of still existing *"jewel steel"*. His report of March 11, 1933 reads shockingly:

"High quality tamahagane is sold out, there are only stocks below the 4th quality level."[76] It should be noted that there are different qualities of tamahagane crude steel as a result of the steel production in the *tatara* furnace, which are briefly listed here: *tsuru ("crane")* = Special Grade, *matsu ("pine")* = First Grade, *take ("bamboo")* = Second Grade, *ume ("plum") A* = Third Grade, *ume ("plum") B* = Fourth Grade, Off Grade, Others.[77]

The reason for these different qualities is the archaic smelting process, which *Edward Hunter* of the *"Department of Arms and Armor, The Metropolitan Museum of Art"* also refers to as a deficiency in his article *"The Japanese Blade, Technology and Manufacture"*: "However, the melting process in the tatara furnace ...is not perfect and tamahagane is full of impurities and does not show an even distribution of the carbon content..."[78] Therefore *Kurata Shichiro* rightly sums up in his report: "It is doubtful that tamahagane of the 4th quality level is suitable for forging good swords."

Usable *tamahagane* for the production of high quality blades is no longer available. *Shichiro's* research on still existing production facilities is just as discouraging: "Since ancient times two modes of operation have been known for a *tatara*. One is called "No-datara" or field-tatara, which was operated on demand at different locations, but it is difficult to find any that still exists today. The other mode of operation is called "semipermanent" and still known production sites are listed in a table. Of these only eight former tatara factories, four are marked "no

---

[76] Tom Kishida, The Yasukuni Swords, Rare Weapons of Japan 1933 – 1945, Kodansha Europe Ltd., 1. Edition 2004, p. 84

[77] Tom Kishida, The Yasukuni Swords, Rare Weapons of Japan 1933 – 1945, Kodansha Europe Ltd., 1. Edition 2004, pp. 90, 92

[78] http://www.metmuseum.org/toah/hd/japb/hd_japb.htm

longer existing", the other four "factory buildings and workers' housing still exist."[79] The inventory ends with the words: "All tatara factories are said to have been closed down at the end of 1925."

Shichiro further complains that there is no technical documentation on how to operate a tatara furnace properly. "Technical knowledge was only passed on orally by the *murage*, men responsible for operating the *tatara*, from generation to generation. This knowledge will be lost when the *murage* have died. There are only a few *murage* left, all of whom are over sixty or seventy years old. It also seems that some of them are too ill to return to their old profession."[80]

Shichiro sees even greater problems in obtaining iron sand in the required quantities. Over the centuries, iron sand has been extracted from rivers *(kawasatetsu)* or in the mountains *(yamasatetsu)* in reasonable quantities according to an old customary law. Although he recognizes that a large-scale (industrial) mining of iron sand poses considerable risks to the ecological systems and the inhabitants of the affected regions, Shichiro argues in his report that the customary law for the extraction of iron sand should be reintroduced as soon as possible.[81]

The supply of charcoal is similarly critical. To smelt 10 tons of iron sand to approx. 2.5 tons of *tamahagane*, 12 tons of charcoal are needed.[82] For the production of charcoal whole forests are cleared, which dramatically changes the landscape. Accordingly, Shichiro's report says: "Almost all the trees in the mountainous region have been felled except those belonging to the

---

[79] Tom Kishida, The Yasukuni Swords, Rare Weapons of Japan 1933 – 1945, Kodansha Europe Ltd., 1. Edition 2004, p. 84
[80] Tom Kishida, The Yasukuni Swords, Rare Weapons of Japan 1933 – 1945, Kodansha Europe Ltd., 1. Edition 2004, p. 85
[81] Tom Kishida, The Yasukuni Swords, Rare Weapons of Japan 1933 – 1945, Kodansha Europe Ltd., 1. Edition 2004, p. 85
[82] http://en.wikipedia.org/wiki/Tatara_(furnace)

Tabe family, and it has become very difficult for the *Torigami Branch Factory* of *Yasukuni Steel Production Company* to obtain charcoal. However, the *Tabe* family has no problem supplying charcoal because they have vast lands and forests in the mountains. In the same way, iron sand can be extracted from their land."[83]

Regardless of the damage to nature and the people who depend on it, massive interventions in the environment and the resulting ecological damage are accepted in order to provide the *Yasukuni-tosho* with the raw material that is indispensable from a puristic point of view, from which swords are to be made to evoke the fighting spirit of the Japanese nation and its soldiers. *Kurata Shichiro* is also pursuing these interests with the help of the media. In his article *(Token Kai Magazine, October 1933)* quoted at the beginning of this chapter, Shichiro argues: "A sword made of "Western Steel" is weak and fragile, unlike the Japanese sword, whose metal has been folded fourteen to seventeen times." With all due respect for the Japanese consciousness of tradition and the pronounced national pride of the Japanese, Shichiro's thesis, with which he tries to explain the superiority of tamahagane over modern industrial steel and perhaps also to justify the overexploitation of nature, cannot remain unchallenged.

Apart from the fact that even swordsmiths who traditionally used western steel folded just as often or more often, Shichiro's thesis can only be explained comprehensibly with the ideologically indoctrinated view of the foundation manager of *Nihon Tanren Kai* and must be considered in the field of propaganda. Shichiro's claim not only denies the fact that imported steel, which was far superior to tamahagane in purity and even distribution of carbon, had already reached Japan from the middle of

---

[83] Tom Kishida, The Yasukuni Swords, Rare Weapons of Japan 1933 – 1945, Kodansha Europe Ltd., 1. Edition 2004, p. 86

the 16th century. He also forgets that important swordsmiths like Minamoto Yoshichika had successfully experimented with western rolled steel and forged *gendaito* in the traditional way, which were not only famous for their sharpness but also for their extraordinary resilience. With his thesis, however, he ignores above all the sad experiences that Japanese soldiers had to make with their samurai swords during the occupation of Manchuria. In fact, many swords were often unable to withstand the extreme temperatures of up to minus 40 degrees Celsius and were severely damaged or broke in battle.

Hikosaburo Kurihara, also known as *Kurihara Akihide*, leading Japanese swordsmith and founder of the *Nihonto Tanren Denshujo (Japanese Swordsmith Institute)*, had gone to Manchuria accompanied by a group of swordsmiths to repair swords and to get a personal impression of the situation on site. In his opinion, the swords found did not meet the expected requirements: "We repaired about 15,000 swords in Shanghai." reported *Hikosaburo Kurihara* after his return. "An officer with a damaged sword who goes into battle the next day is a merciful sight. I saw how many of them worked late into the night on their swords, which meant life or death for them." And further: "Blades of good steel do not break as easily as those we found. I have therefore recommended to the War Ministry that all swordsmiths in the country be supplied with Manchurian steel *(mantetsu)*. This steel is stronger than any other."[84]

---

[84] Richard Stein, Japanese Sword Guide,
http://www.japaneseswordindex.com/koa.htm

Japanese infantry column on the march through the Manchurian winter.

The swords forged from steel that was "stronger than any other" were *Mantetsu-to*. Contrary to the longstanding abuse and disparagement by self-proclaimed Western "sword experts" that *Mantetsu-to* were forged from "old railroad tracks", the only thing that links these swords to the railroad is the fact that their steel was developed in the research department of the *South Manchuria Railway Company*. The *South Manchuria Railway Company* operated within the Japan-controlled zone after the occupation of Manchuria by Japan. Its business activities included the operation of the railroad and railroad lines, coal mining and research. In early September 1937, the *South Manchuria Railway Company* had already begun to develop a steel that was exclusively reserved for the production of sword blades.[85]

A factory especially for the production of swords was built and in November 1937 the production started. The training of the swordsmiths in the factory was carried out by the swordsmiths *Takeshima Hisakatsu* and *Wakabayashi Shigetsugu*. First these swords were signed *"Mantetsu Kitau Tsukuru Kore"*. In 1939, the monthly production rate increased to 400 swords and the patriotic phrase *"Koa Isshin" ("Asia one heart")* was used for the signature.[86] *Mantetsu-to* were manufactured with great care. This also applies to the tang *(nakago)* and the calligraphically carefully executed signature. The *hada* shows densely forged *ko-itame*, the *hamon* is usually executed in *suguha*. In 1944 an exhibition of *shinsaku-to ("Newly Forged Swords")* was held at *Yasukuni Shrine* under the auspices of the Imperial Army. Besides the *Yasukuni-to*, another section was dedicated to the *Mantetsu-to*, which clearly proves their importance. Such a sword was presented to *Lieutenant Colonel A.K. Crookshank* in *Bentong, Malaya*, by *Major General Tamoto* on the occasion of the Japanese surrender on September 2, 1945, and is proof that even high-ranking officers of the Imperial Army proudly wore *Mantetsu-to*. The sword is now in the National Army Museum in London. Due to their robustness and extreme resilience, *Mantetsu-to* are also very popular among swordsmen.[87]

Also *Nakayama Hakudo* was an advocate of the new technologies due to his extensive practical experience as long as the swords were traditionally forged and proved this again and again by cutting tests. On July 10, 1934, *Nakayama Hakudo* demonstrated the strength of the new blades in a public demon-

---

[85] South Manchuria Railway Co.Ltd. Dalian Railway Factory Sword Works, "Kōa Issin", Published 1939
[86] Richard Stein, Japanese Sword Guide, http://home.earthlink.net/~steinrl/koa.htm
[87] http://de.wikipedia.org/wiki/Gunto

stration by cutting through a finger-thick iron bar with a single stroke. The rod was wrapped with rice straw soaked in water, as is customary in cutting tests. The sword had been forged at *Nihonto Tanren Denshujo* by students of *Hikosaburo Kurihara*.[88]

But even though *tamahagane* was outdated from a metallurgical point of view due to the impurities and the uneven distribution of the carbon content compared to modern industrial steels, Japan's "Jewel Steel" may also have peculiar advantages precisely because of these disadvantages. Not least of all, it is this very fact that may give the blades the texture that makes the steel used unmistakable and by which we can, for example, infer its origin or recognize periods in which a sword was forged. The texture of a blade from *tamahagane*, for example, is different from the texture of a blade from *namban-tetsu*, just as a blade whose steel is made from iron sand from *Bungo* province is darker in color than a blade whose steel comes from *Sendai*. Thus, among other factors, texture is ultimately an important indicator for the evaluation and classification of a blade, just as it is in other art forms, for example, in painting, where the background or the application of paint are important indicators of an era or an artist. No matter whether *tamahagane* or *namban-tetsu,* in the end, the only thing that matters with a blade that has been manufactured in a traditional way is the individual craftsmanship and artistic characteristics inherent in it. Because only then is it decided whether a sword is only a weapon or also a work of art.

After this digression on the usability and properties of sword steels, let us come back to the *Yasukuni-to*, which were produced at the shrine using the best *tamahagane* and maintaining high quality standards. This also included cutting tests and sword inspections, which took place regularly at the shrine.

---

[88] http://en.wikipedia.org/wiki/Nakayama_Hakudo

Swords were randomly selected from a number of blades that had already received their basic polish and a cutting test was carried out with every third sword.[89] Likewise, the swords forged at the shrine were thoroughly inspected twice a month. The inspected swords were divided into four classes according to strict criteria, whereby the fourth class meant that the sword had not met the requirements *("disqualified")*.[90]

Yasukuni-to were created in the style of the blades of *Nagamitsu* and *Kagemitsu,* which belonged to the *Osafune School*. This school worked during the *Kamakura Period* (1185 - 1333) in the *Bizen* province. With their *ko-kissaki,* the densely forged *ko-itame hada,* the skilfully applied *suguha hamon* and the deep *bizen-sori,* the elegantly tapering blades look like the counterparts of later *Koto-tachi*. This impression is further enhanced by the special texture of the blade surface *(ji-tetsu),* which was significantly influenced by the properties of the available *tamahagane.*

---

[89] Tom Kishida, The Yasukuni Swords, Rare Weapons of Japan 1933 – 1945, Kodansha Europe Ltd., 1. Edition 2004, p. 72
[90] Tom Kishida, The Yasukuni Swords, Rare Weapons of Japan 1933 – 1945, Kodansha Europe Ltd., 1. Edition 2004, p. 73

Yasuoki

One of these swords, which were forged by *Shimazaki Yasuoki* at *Yasukuni Shrine* and which was intended as a gift of honor for a shrine, is now presented in further detail. This sword is a special *tachi*, which Yasuoki forged in 1940 by order of the governor of *Kanagawa* province. The sword is documented in the book *"The Yasukuni Swords Rare Weapons of Japan 1933 -1945"* by *Tom Kishida* in the *"Swordmaking Record of Yasuoki"* shown on page 58. This is *Work No. 13* in the work report hand-written by Yasuoki for the month of August 1940.[91]

The sword is signed like all Yasukuni-to *tachi-mei* and is designed in the style of the works of *Nagamitsu* and *Kagemitsu*. The *tachi* in its classical form with *shinogi-zukuri, ihori-mune, deeper bizen-sori, ko-kissaki* and the proportionally tapering blade *(motohaba 29 mm, sakihaba 19 mm, motokasane 7 mm, sakikasane 5 mm)* reminds of a late koto blade *(sue koto)*. The *hada* shows dense and absolutely flawless forged *ko-itame*, the *hamon* is designed as a *suguha hamon* and approaches the edge in a slight arc below the *yokote*. This arc is a clear identification mark and typical for the early blades of the *Yasunori Group*.[92]

The tang is aesthetically designed as *kiji-momo nakago*[93], the file marks are very carefully worked *kiri-yasurime*. Yasuoki made *kiji-momo nakago* up to his 349th sword at the beginning of August 1942, after which he changed the shape of the tang

---

[91] Tom Kishida, The Yasukuni Swords, Rare Weapons of Japan 1933 – 1945, Kodansha Europe Ltd., 1. Edition 2004, p. 58
[92] Tom Kishida, The Yasukuni Swords, Rare Weapons of Japan 1933 – 1945, Kodansha Europe Ltd., 1. Edition 2004, p. 55
[93] Tom Kishida, The Yasukuni Swords, Rare Weapons of Japan 1933 – 1945, Kodansha Europe Ltd., 1. Edition 2004, p. 55

as well as the file marks on the back of the tang.[94] The tang is signed on the omote "Yasuoki" and on the ura "showa 15 nen 8 gatsu kichi jitsu" ("Showa 15th year 8th month (August 1940) A lucky day"). The signature comes from Yasuoki himself *(jishin-saku)* and not from one of his *sakite* (assistants). These swords were known as *dai-saku*. See also page 59, *"The Yasukuni Swords"* by *Tom Kishida,* where the difference between the signatures *jishin-saku* and *dai-saku* is explained in detail.[95] Similarly, the character for *"oki"* is the same as the older version with which Yasuoki signed his early blades before changing this character when he completed his 181st sword, which he forged in early September 1941.[96]

In his 1940 work report, Yasuoki indicated the length of the blade *"No. 13"* as *1 shaku, 9 sun, 6 bu.* In fact, Yasuoki's present blade measures from the *mune machi* to the tip of the blade just a hair less than 594.0 mm *(1 shaku, 9 sun, 6 bu = 593.88 mm).* This distinguishes this *tachi* from the other *Yasukuni-to,* which normally have a length of more than *two shaku,* at first glance, because the average blade length *(nagasa)* of the swords forged at the shrine was about 66.7 cm.

But what also makes this non-standard sword a special and absolutely unique *Yasukuni-to* are the handwritten notes that Yasuoki wrote behind his "Work N. 13" in his personal work report. Translated, the Japanese characters mean: "To commemorate the 2600th anniversary of the Japanese Empire, the governor of Kanagawa province throughout Japan has commis-

---

[94] Tom Kishida, The Yasukuni Swords, Rare Weapons of Japan 1933 – 1945, Kodansha Europe Ltd., 1. Edition 2004, p. 59
[95] Tom Kishida, The Yasukuni Swords, Rare Weapons of Japan 1933 – 1945, Kodansha Europe Ltd., 1. Edition 2004, p. 59
[96] Tom Kishida, The Yasukuni Swords, Rare Weapons of Japan 1933 – 1945, Kodansha Europe Ltd., 1. Edition 2004, p. 59

sioned ten master smiths to forge ten swords as gifts for ten shrines in Kanagawa."[97]

Taking all the facts into account, there is no doubt that this *tachi*, which was forged at *Yasukuni Shrine* by *Yasuoki*, is one of the ten swords that the governor of *Kanagawa* commissioned from ten master swordsmiths in 1940 to present to ten shrines on the occasion of the 2600th anniversary of the Japanese Empire. One of these ten master swordsmiths selected by the governor all over Japan was the *Yasukuni-tosho Shimazaki Yasuoki*. Because of this history, personally written down by Yasuoki in his work report, this tachi is also of outstanding cultural and historical importance and represents a unique legacy of the *Yasukuni shrine* and its swordsmith *Shimazaki Yasuoki*.

The *tachi* shows all the beauties you can expect from a Yasuoki blade and testifies until today to the high craftsmanship and artistic skills of *Yasukuni-tosho* and the outstanding quality of the swords traditionally forged at *Yasukuni Shrine*. *Tom Kishida* rightly states: „The 8100 swords forged at Yasukuni Shrine between 1933 and 1945 represent a special legacy of artwork in which not only time-honored forging methods live on, but also the aesthetics and spirit of the samurai warriors."

---

[97] Tom Kishida, The Yasukuni Swords, Rare Weapons of Japan 1933 – 1945, Kodansha Europe Ltd., 1. Edition 2004, p. 58

One of 10: A culturally and historically significant Yasukuni-to and unique legacy of the Yasukuni-tosho Shimazaki Yasuoki. Yasuoki forged the sword in August 1940 by order of the Governor of Kanagawa Province on the occasion of the 2600th anniversary of the Japanese Empire. To commemorate this memorable year, the governor commissioned ten master swordsmiths throughout Japan to forge ten swords intended as gifts for ten shrines in Kanagawa. One of these ten smiths was the Yasukuni-tosho Shimazaki Yasuoki and the sword presented here is one of these ten swords.

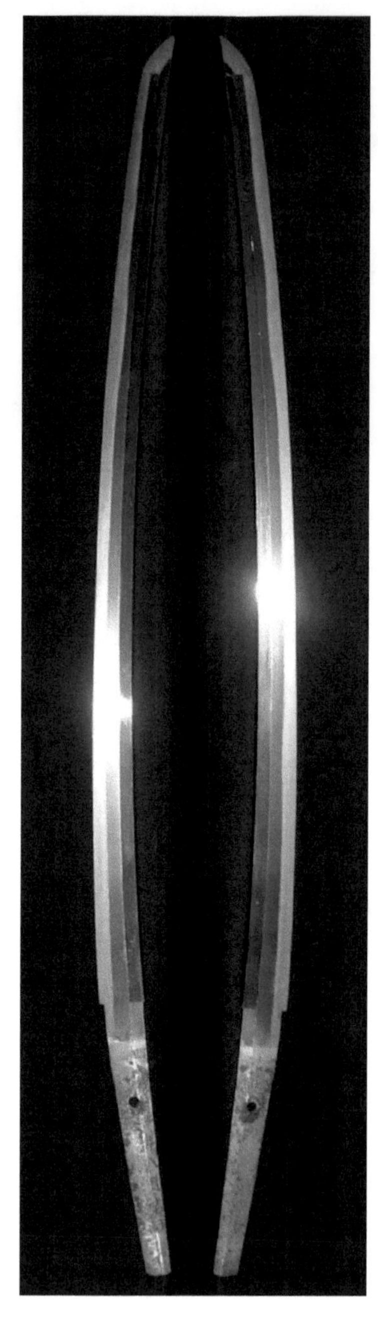

98

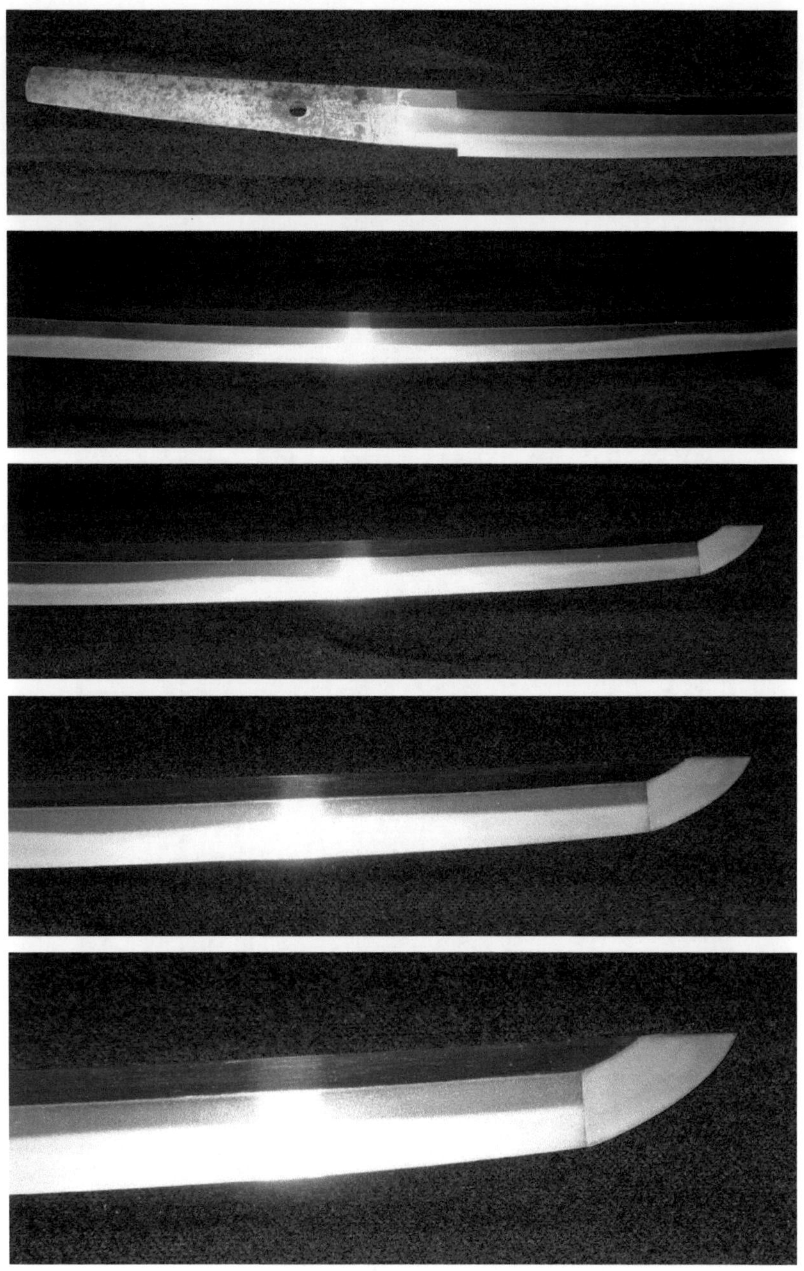

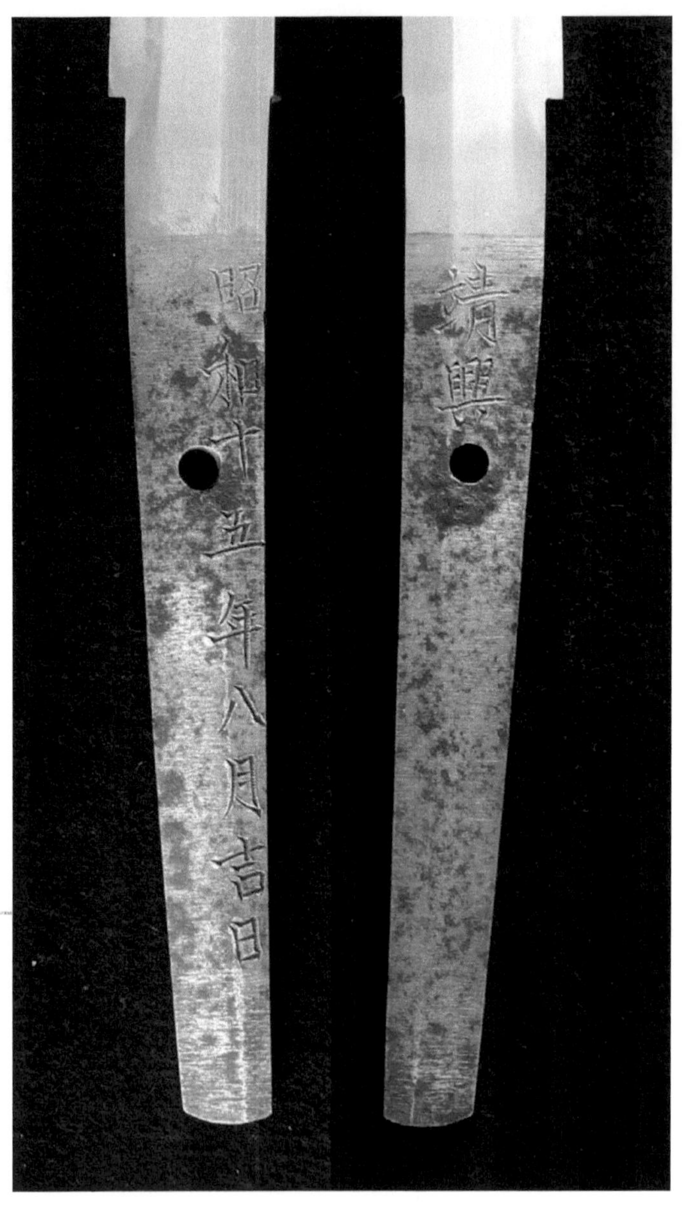

The sword tachi-mei "Yasuoki", the ura dated "Showa 15th year 8th month (August 1940) A lucky day".

Gunto, its Place in the Imperial Army and its Importance in Battle

"The sword is the first weapon in the history of mankind whose origin is not to be found in one of the tools for daily use, such as axe and knife, nor in one of the hunting tools once essential in the fight for existence, such as spear or bow and arrow, but which from the beginning was exclusively intended for killing people. Probably for this macabre reason, the sword is generally regarded as the noblest of all weapons, and so a separate myth about the sword ...has formed among all peoples and cultures who know it."[98]

In Japan this myth is particularly pronounced. According to legend, Susanoo, brother of the Sun Queen Amaterasu, slayed the eight-headed dragon Orochi with a sword. When he cut off one of the dragon's eight tails, a sword blade of unbelievable beauty jumped out, which he presented to Amaterasu to mark his victory. After Susanoo's descendants had taken over the rule on earth, Amaterasu sent her grandson Ninigi to plant rice and rule the earth. The union of a descendant of Ninigi with a daughter of the Dragon King resulted in Jimmu, the first Japanese Tenno and thus the ancestor of all Japanese emperors. According to tradition, Amaterasu gave Jimmu three gifts, which are still the throne insignia *(sanshu no jingi)* of the Japanese imperial family.[99] Along with the necklace and the mirror with which the gods once lured Amaterasu out of the cave in which she had been hiding, this was the sword Susanoo had won in battle with the dragon Orochi.

---

[98] Helmut Nickel, Curator of Arms & Armor, Metropolitan Museum of Art, New York, "The Japanese Blade: Technology and Manufacture", http://www.metmuseum.org/toah/hd/japb/hd_japb.htm
[99] http://de.wikipedia.org/wiki/Ninigi

In Japanese mythology, the sword is the gift of the gods and the embodyment of divinely bestowed power. As a sign of its status, the court nobility *(kuge)* already wore magnificently mounted *tachi* in early times. With the growing importance of the warrior nobility *(buke)* the sword gains also for the Samurai besides its practical meaning as a deadly weapon more and more in importance as a visible status symbol and expression of their social status. Since the Shogunate took power it was the exclusive privilege of the samurai to carry two swords, which they knew how to handle perfectly through intensive mental and physical training.

The Japanese showed obsequious veneration to the samurai, and not only because the samurai belonged to one of the highest ranks in Japanese society. It was the code of honor of the samurai that established their moral integrity and to which the members of all classes always paid deep respect. This code of honor raised seven virtues to the maxim of their actions and made the *"Law of Bushido"* immortal: *Gi* (義): sincerity and justice, *Yu* (勇): courage, *Jin* (仁): kindness, *Rei* (礼): politeness, *Makoto* (誠) or *Shin* (真): truth and truthfulness, *Meiyo* (名誉): honor and *Chugi* (忠義): faithfulness or also *Chu* (忠): duty and loyalty.[100] Absolute loyalty and the willingness to lay down one's life for one's liege lord at any time without hesitation and in faithful fulfillment of duty determined the ideal of life and the path of the samurai.

A strict and extremely hard upbringing, which began in early boyhood, laid the foundation for the extraordinary hardness and cold-bloodedness of this warrior elite. Early on the boys were accustomed to privation, hunger and cold. They had to walk barefoot even in the harsh winter or were abandoned in bitter cold in places unknown to them, from where they had to

---

[100] Gewitsch, Michael, „Die sieben Tugenden der Samurai",
Ausarbeitung zum 1. DAN Jiujitsu, 2013

fight their way home on their own. Occasional food deprivation accustomed them early on to endure hunger, for it was shameful for a samurai to feel or express hunger. The boys had to attend executions, stay overnight at the places of execution or return there at night and leave signs at the place of execution or the executed to prove that they were actually there. Accustomed to the horror and the sight of death in early youth, death lost its horror for them.[101]

Loyalty and truthfulness, courage and justice, kindness and politeness were among the virtues of the samurai as well as compassion. However, the compassion of a samurai knew macabre limits. If a member of a lower class did not show the necessary respect to a samurai, the samurai had the right to "strike down the disrespectful one and leave".[102] Life and death were close together in feudal Japan and did not mean much to the samurai. The samurai lived with death and was always ready to die for honor and for his master. This went so far that faithful samurai often committed ritual suicide *(seppuku)* at the death of their master to follow their master to the other world. This tradition lived on into the 20th century. When *Emperor Meiji* died in 1912, his faithful general *Nogi Maresuke* followed him into death.

General Nogi Maresuke was born on December 25, 1849 in Edo, today's Tokyo. "During the Japanese civil war of 1877, he served as a captain in the Imperial Army. Later, he commanded a brigade in the Chinese-Japanese War (1894-1895) and was appointed Lieutenant General in 1895. His greatest military achievement is considered to be the 154-day siege of Port Arthur during the Russo-Japanese War (1904-1905). In the meantime promoted to general, he forced the Russian troops to surrender on January 2, 1905. On the one hand, this was a great

---

[101] Mauer, Kuno, Die Samurai, Econ Verlag Düsseldorf, 1. Auflage 1981
[102] Tsunetomo Yamamoto, Hagakure, Kabel-Verlag, 7. Auflage, März 2009

success for the Japanese armed forces. However, the victory claimed 58,000 lives on the Japanese side, including Nogi's own two sons. The losses of the Russian troops amounted to 31,000 dead. On September 13, 1912, General Nogi Maresuke and his wife Shizuko committed seppuku to follow the late *Emperor Mutsuhito,* better known as *Meiji Tenno*, to his death. General Nogi Maresuke and his wife are still worshipped in various Nogi shrines as *kami ("revered spiritual beings")*.[103]

Emperor Mutsuhito (left in the picture) and his loyal general Nogi Maresuke.

Tsunetomo Yamamoto, the author of *"Hagakure"*, was also a samurai. After the death of his master he was not allowed to follow him, because his master had forbidden him by decree to commit *seppuku*. So he became a *Zen monk* and between 1710 and 1716 he wrote his still famous work *"Hagakure"*. In it it says: "I have found out bushido, the way of the warrior lies in dying. If one is confronted with two alternatives, life or death,

---

[103] http://de.wikipedia.org/wiki/Nogi_Maresuke

one should choose death without hesitation. There is nothing difficult about that; you just have to be determined to pursue your goal." and further "The way of the samurai is death..."[104] The basic Buddhist attitude that death is a welcome conclusion to life on earth, the death-deprecating education and their strong code of honor may have been the reasons for the extraordinary bravery and success of the samurai in battle, which was always proven. They were famous for this, just like their swords, with which they became masters of life and death. There are countless stories about the heroic deeds of samurai warriors. One of them is the famous story of the *"47 Ronin"*, which every child in Japan, and not only there, knows and which still commands the greatest respect and approval from people with a strong sense of honor and loyalty. Due to the unconditional integrity and loyalty of the samurai and their unique role model function in Japanese society, it becomes understandable why the soldiers of the Imperial Army strove to carry a sword that embodied the high ideals of the samurai and the spirit of bushido for them and to which they could entrust their life and soul in battle.

Mori Masahiro, Japanese author, writes in his essay "SOLDIERS AND GUNTO"[105]: "In war reports and the propaganda of the Imperial Army there are many stories about the Japanese sword, which suggest its superiority over modern firearms. These stories often exaggerate the effectiveness of the Japanese sword by telling heroic deeds of army officers who wield their swords in battle. These stories are ubiquitous. More important than the credibility of such stories, however, is the fact that these stories prove that the Japanese sword had become a powerful symbol of loyalty and militarism by mentally bringing together the spirit of bushido and the Imperial soldiers."

---

[104] Tsunetomo Yamamoto, Hagakure, Kabel-Verlag, 7. Auflage, März 2009
[105] Tom Kishida, The Yasukuni Swords, Rare Weapons of Japan 1933 – 1945, Kodansha Europe Ltd., 1. Edition 2004, p. 104

"The Imperial soldiers instinctively felt that there was a deep and important meaning to carrying a samurai sword. Throughout its long history it had been revered and treated as the soul of the samurai. Beyond its importance as a weapon, it had a deep symbolic power. It possessed mystical abilities, not only protecting the life of the warrior, but also helping him to put into practice bushido, whose basic principles teach to overcome any fear of death. Mindful of his own weakness and fragility, the warrior personified his weapon, which was imbued with soul and character."

"During the Second World War, pilots took their swords into the machines and wounded soldiers did not part with their swords even in the military hospital. It is reported that soldiers were encouraged to look at their swords before a final, deadly attack to sharpen their senses, and high-ranking officers always took care of their swords personally and with dedication." "The Japanese sword has always been collected and studied by people of the upper class. However, the soldiers of the Imperial Army were neither samurai nor members of the nobility. It is beyond our imagination how proud they must have been to carry a sword at their side in an era that still had features of a feudal hierarchy."

"The soldiers of the Imperial Army of World War II and the bushi of old Japan should not be lumped together, because these two classes of warriors have different historical backgrounds. However, if people pay respect to swords that were carried and used by warriors in the Sengoku and Bakumatsu periods, then gunto must also be re-evaluated and seen as the soul of the soldiers who used them. Furthermore, we must not ignore the fact that gunto has proven its effectiveness as a weapon in war, case by case." As far as Mori Masahiro in his article "SOLDIERS AND GUNTO".

Gunto, its Place in the Imperial Army
and its Importance in Battle
Picture Section

The following image documents correspond to Mori Masahiro's theses. They come from press or state archives and photo albums of Japanese soldiers and bear witness to the high status the samurai sword had in Japan's Imperial Army until the end of World War II. They show the Sun Emperor with his Samurai sword in gunto mounting in an almost mystical transfiguration on his stallion Shirayuki and the young lieutenant, who proudly and no less symbolically poses for the photographer with his gunto, but also convey oppressive impressions from the hell of the material battles through which the soldiers of all warring parties passed.

The pictures tell of the brutal force with which the war hit people and material. They only give us an idea of the enormous forces that wiped out entire cities, brought down airplanes from the sky, sent heavily armored battleships to the bottom of the sea, and transformed just recently flourishing areas under the fire of artillery into desolate crater landscapes. We know that the swords could not stand against this destructive force. The author himself has seen swords severely damaged or shot by shrapnel and shell splinters. Nevertheless, soldiers of all branches of service firmly believed in the spiritual power of their swords and entrusted their lives to them in the inferno of the material battle.

We know that the swords did not only accompany the soldiers of the army into battle; pilots also did not go on enemy flight without their swords, just like the officers of the Imperial Navy who took their swords on board when they went on enemy ride. Unlike the swords of the land forces, these swords were rarely or never used in practice. And although they remain hidden from our eyes in the photos, they accompanied their proud bearers in the cockpits of the airplanes and on board the ships and helped them to overcome the fear of death and to follow the path of the warrior with determination.

The picture section also recalls the "Tiger of Malaysia", General Yamashita Tomoyuki, for whom the path of the warrior was tragically fulfilled with the sentence of death by hanging. The composure with which he climbed the gallows deserves our utmost respect. Without fear of death and in the conviction not to be ashamed before the gods, he asks them to bless his executioners. The events of the war and the verdict by a U.S. military tribunal reveal the double standards and the arbitrariness of the victor, who finds the defeated guilty despite considerable doubts about his guilt in order to make an example, while the victor, without need and unpunished to this day, killed about a quarter of a million civilians by dropping two atomic bombs on Hiroshima and Nagasaki, who either died directly in the drops or suffered and died horribly from the atomic after-effects.[106]

---

[106]https://de.statista.com/statistik/daten/studie/1086264/umfrage/geschaetzte -zivile-todesopfer-und-verletzte-in-hiroshima-und-nagasaki/

Emperor Hirohito on his stallion Shirayuki (White Snow): The Tenno carries a samurai sword in shin-gunto mounting. The pose symbolizes in an almost mystical transfiguration the fusion of the Divine Sun Emperor with the spirit of Bushido. The radiant power and the effect on the soldiers of the Imperial Army and the Japanese people become comprehensible to people from other cultures when they take a closer look at the subject.

A momentous visit: **Emperor Hirohito** *accompanied by* **Hajime Sugiyama** *(on the left behind the Tenno) and* **Nara Takeji** *(on*

the right behind the Tenno) visiting the Yasukuni Shrine. The emperor and his companions wear uniforms and their gunto. Shortly after this visit, Japan invaded China (Second Japanese-Chinese War from July 7, 1937 to September 9, 1945). This photo comes from the photo album of a Japanese soldier.

Hajime Sugiyama was born on January 1, 1880 as a scion of a samurai family in Kokura, Fukuoka Prefecture, and had a steep military and political career. In the Imperial Army, he rose to the rank of Gensui (Field Marshal) and was Minister of the Army several times. As Minister of the Army, he was responsible for the outbreak and escalation of the Second Japanese-Chinese War (incident at Marco Polo Bridge).[107] In December 1938, he received the supreme command of the regional army in northern China. After his return to Japan, he was initially the head of Yasukuni Shrine. In September 1940, he was appointed chief of the Army General Staff, and in this function he strongly advocated a preventive strike against the United States. "He promised Japan a quick victory in the event of war, but on September 5, 1941, only two months before the outbreak of the Pacific War, he was scolded by Emperor Hirohito for having promised to defeat China within three months as Army Minister in 1937. In this context, the Tenno questioned his confidence in a quick victory over the Western powers."[108] After Japan's surrender on September 2, 1945, he pushed ahead with the demobilization of the troops under his command, as demanded by the Allies. Then he put an end to his life on September 12 by shooting himself in the chest four times with a revolver at his desk. In their house, his wife also committed suicide. His grave is in the Tama cemetery in Fuchu, Tokyo.

Nara Takeji was born on April 28, 1868 near the present town of Kanuma in Tochigi Prefecture as the son of a farming family.

---

[107] http://de.wikipedia.org/wiki/Zwischenfall_an_der_Marco-Polo-Brücke
[108] http://de.wikipedia.org/wiki/Sugiyama_Hajime

He attended the Japanese Military Academy and the Army Artillery School. From 1894 to 1895, he took part in the First Japanese-Chinese War and successfully continued his military career after his return. He served in the general staff of the Imperial Army and was sent to Germany as a military attaché. During the Pan-Japanese Russian War, Nara was commander of the Independent Heavy Artillery Brigade. After the war and another visit to Germany, he was promoted to deputy minister of war. In 1918 he participated in the negotiations for the Treaty of Versailles as a representative of the Japanese delegation. Subsequently, Nara became Personal Adjutant to Crown Prince Hirohito and, after his coronation as Emperor, was appointed Chief Adjutant to His Majesty the Emperor. He supervised the Crown Prince's training in military affairs in theory and practice. In 1921 he was part of Hirohito's entourage on his official European tour. In 1924 he was promoted to general, and from 1933 he belonged to the Emperor's "Privy Council".[109] After his retirement from military service he was raised to the rank of a baron. In postwar Japan he was president (chairman) of the "Dai Nippon Butoku Kai".[110] Baron Nara Takeji died on December 21, 1962, and his diaries from his time as the emperor's personal adjutant, which had been well kept until then, provided posterity with previously unknown insights into the thoughts and role of the Tenno during the Second World War.[111]

---

[109] http://en.wikipedia.org/wiki/Privy_Council_Japan

[110] http://en.wikipedia.org/wiki/Dai_Nippon_Butoku_Kai

[111] http://en.wikipedia.org/wiki/Takeji_Nara

"The sword is the soul of the samurai." On the left, Emperor Hirohito with a samurai sword in a magnificent gala uniform. No less proud, a lieutenant of the Imperial Army with his gunto poses for the photographer before he goes to war for Emperor and Empire. In battle, the officers entrusted their lives to their swords, just as the soldiers trusted their officers and followed them when the "last samurai" with the sword in their hands led their men to storm the enemy positions.

Pearl Harbor, Hawaii: On the morning of December 7, 1941, Japanese naval aviators attack the US Pacific fleet anchored in Pearl Harbor. During the attack, a large part of the fleet is destroyed or in some cases severely damaged. At the same time, the Japanese offensive against the British and Dutch colonies in Southeast Asia begins. Thus the war in Europe expands into a globally led world war. The attack is regarded as a decisive turning point for the entire course of the war, because the USA, which had been neutral until then, declared war on Japan and for the first time actively intervened in the war events of the Second World War. The picture comes from the archives of the Imperial Japanese Navy and shows a Japanese fighter bomber attacking Pearl Harbor.[112]

---

[112] https://en.wikipedia.org/wiki/Attack_on_Pearl_Harbor

Pearl Harbor December 7, 1941, 08:06 am local time: The "USS Arizona" is hit by an armor-piercing 800-kg bomb which penetrates the upper deck between the forward gun turrets and explodes the two ammunition chambers. The floating colossus sinks in only nine minutes. The picture documents the tremendous force of the explosion, which killed 1,177 men. Only 289 crew members survived the attack. One of them was Seaman 1C James Daniel Lancaster, who later described his impressions of the devastating attack: "I had never seen so many planes in the air before. When I saw the rising sun under the wings, I knew what was in store for us." As he ran to his battle station, a bomb exploded amidships. Gasping for air, he regained consciousness in the water. "The water was on fire. The Arizona was a huge ball of fire. All around me were men screaming, a scream I will never forget." Lancaster dived away under the burning water and reached the captain's boat. Although badly wounded himself, he tried to reach the swimmers around him. After taking ten men to Ford Island, Lancaster returned to the inferno for a second rescue operation. Once

again he managed to pull some of his comrades out of the water. He was awarded the "Purple Heart" for his bravery.[113]

The "USS Arizona" on a photo from 1931. The "USS Arizona" was a Pennsylvania-class battleship and one of the most powerful ships of the time. Her sinking on December 07, 1941 became a symbol of American humiliation. On December 29, 1941, the wreck was registered as a war grave. The "USS Arizona" lies at a depth of 12 meters and is still the final resting place for 1,102 crew members.[114]

---

[113] http://www.ussarizona.org/stories/uss-arizona-survivor-stories/107-lancaster-james-daniel-usn

[114] https://de.wikipedia.org/wiki/USS_Arizona_(BB-39)

August 24, 1942, Battle of the Eastern Solomon Islands: A Japanese aircraft bomb hits the flight deck of the "USS Enterprise". Photographer Robert Frederick Read loses his life in this photograph. The aircraft carrier received a total of three bomb hits in which 77 sailors were killed and 91 wounded. The carrier gained tragic fame when, after the Japanese attack on Pearl Harbor, it arrived in Pearl Harbor a day later and sent planes ahead to assess the damage in the harbor. Eighteen of the planes over Pearl Harbor were greeted by the fire of their own troops, as it was believed that the Japanese would attack again. Six planes were shot down and eight crew members were killed. [115]

---

[115] https://de.wikipedia.org/wiki/USS_Enterprise_(CV-6)

The "USS Enterprise" on an official photo of the U.S. Navy from April 12, 1939.

July 28, 1942, Kokoda, Papua New Guinea: Lieutenant Colonel Hatsuo Tsukamoto gives the order to attack and leads the Japanese Marines (5th Sasebo Special Naval Landing Forces) with a drawn samurai sword to storm the Kokoda airfield and adjacent town.[116] *In July 1942 the Japanese had landed at Basabua on Papua New Guinea. The aim was to advance by land along the grueling Kokoda Trail to Port Moresby and take the strategically important port. The changeable battles for Kokoda took place in late July to early August 1942, with the Japanese facing Australian troops on the Allied side, supported by the United States.*[117]

[116]https://commons.wikimedia.org/w/index.php?search=Hatsuo+Tsukamoto+&title=Special:MediaSearch&go=Go&type=image
[117]https://en.wikipedia.org/wiki/Battle_of_Kokoda

Samurai swords in the material battle: Japanese soldiers attack an American Stuart tank (official designation "Light Tank M3") with infantry means. In the foreground on the left of the picture, an officer with a drawn samurai sword awaits the exit of the tank crew together with his men. The photo was taken in 1942 during the Pacific War on the Philippines. The Pacific War began with the outbreak of the Second Japanese-Chinese War on July 7, 1937. After the attack on Pearl Harbor on December 7, 1941, the U.S. entered this conflict the following day and thus the Second World War against the Axis Powers (Germany, Italy, Japan). As Western allies, Great Britain, Australia, New Zealand and the Netherlands fought alongside the Americans in the Pacific. [118]

---

[118] https://de.wikipedia.org/wiki/Pazifikkrieg

Spring 1942, Bataan, Central Region Luzon, Philippines: Under the command of an officer (second from the right with a drawn samurai sword), Japanese infantry uses a flamethrower to attack an enemy bunker position. On December 10, 1941, the Japanese had begun the invasion of the Philippines on Luzon Island. The Allied American and Filipino units stationed there under the command of General Douglas MacArthur were far inferior to the Japanese. Already on the first day, Japanese planes succeeded in taking out most of the American planes on the ground and gaining air supremacy. This enabled the Japanese to bring their ground troops ashore almost unhindered. MacArthur decided the orderly withdrawal of all units to the Bataan peninsula, where they fought bitterly for every foot of ground until the surrender of the Allies.[119]

---

[119] http://de.wikipedia.org/wiki/Pazifikkrieg

April 9, 1942, Bataan, Philippines: After the capitulation of the allied troops, officers of the Imperial Japanese Army jubilantly raise their samurai swords into the air. After the surrender, some 70,000 men were taken prisoner of war by the Japanese. After the capture, the death march from Bataan began, in which the prisoners had to walk from the south of the peninsula

to a train station about 100 km away. About 16,000 Allied soldiers lost their lives on the march.[120]

Group picture with swords: Japanese officers after landing on Bougainville, Papua New Guinea. In March 1942, the Japanese occupied the island and set up several airfields to secure their strategically important naval base in Rabaul, Papua New Guinea, 300 km away. Unexpected by the Japanese, in November 1943 landed the first allied troops near Cape Torokina. Due to the Allied sea and air superiority, the Japanese were subsequently under permanent air attacks and were cut off from supplies. Despite this, they continue to resist fiercely in the jungle for another two years. On September 8, 1945, about three weeks after Japan's official surrender, the last Japanese soldiers went into captivity on the island. During the fighting, 18,500 Japanese were killed. The Allied losses are estimated at 1,100 casualties.[121]

---

[120] http://de.wikipedia.org/wiki/Todesmarsch_von_Bataan
[121] https://de.wikipedia.org/wiki/Bougainville

October 26, 1942, Santa Cruz Islands: The "USS Hornet" is attacked by Japanese naval aircrafts. The photo shows a Japanese naval aviator swooping down on the "USS Hornet" in a dive.

On October 26, 1942, the "USS Hornet" sailed with escort ships to the Santa Cruz Islands to intercept the Japanese fleet. After successful reconnaissance of the Japanese unit, its 54 carrier aircrafts took off in two waves. The Japanese, for their part, attacked the American unit with torpedo and dive bombers. Warnings reached the "USS Hornet" too late. At 09:10 am local time the carrier received a first bomb hit on the starboard side of the flight deck. Three minutes later, a crashing Japanese aircraft crashed through the flight deck. The plane still had three bombs on board, two of which detonated on impact. Seven minutes later, another Japanese aircraft crashed into the forward gun battery and exploded. Towing attempts by the heavy cruiser Northampton failed because of the continuous attacks of the Japanese. Realizing that the ship could not be saved, the escort destroyers Mustin and Anderson were ordered to sink the Hornet. They fired nine torpedoes and about 300 shells of 127 mm caliber at the Hornet. However, this was not enough for a fast sinking of the carrier. The Hornet was then abandoned to its fate. In the end, the Japanese destroyers Makigumo and Akigumo with four torpedoes gave it the catching shot. The USS Hornet sank to a depth of 5000 meters off the Santa Cruz Islands on October 27, 1942 at 1:35 am local time. Most of the crew could be saved by the escort ships; nevertheless 111 men died, 108 were wounded. The "USS Hornet" belonged to the Yorktown class like the "USS Enterprise" and the "USS Yorktown". The three carriers were put into service between 1937 and 1941 and together with the two Lexington-class aircraft carriers ("USS Saratoga" and "USS Lexington") formed the backbone of the American carrier fleet at the beginning of World War II. The Yorktown was sunk in the Battle of Midway, the Hornet in the Battle of Santa Cruz. Only the „USS Enterprise" survived the war.[122]

---

[122] http://de.wikipedia.org/wiki/USS_Hornet_(CV-8)

October 26, 1942, the fateful day of the "USS Hornet": The gun crews on the 40 mm Bofors twin guns fire what the barrel gives. Despite this, the Japanese pilots, who were determined to do everything, succeeded in breaking through the raging barrage of the ship's flak and fulfilled their mission.

"The way of the warrior is to die." In the massive defensive fire of the ship's flak, the fate of a Japanese naval aviator is fulfilled.

General Yamashita Tomoyuki, the "Tiger of Malaysia", with his samurai sword in shin-gunto mounting.[123]

---

[123] https://en.wikipedia.org/wiki/Tomoyuki_Yamashita

Yamashita Tomoyuki was born on November 8, 1885 in Osugi on the island of Shikoku, Japan. He completed his military training in 1906 at the military academy of Hiroshima with the rank of a lieutenant. First combat mission in 1914 during the siege of Tsingtau, China. As an expert on the German-speaking world, Yamashita worked from 1919 to 1922 as deputy military attaché, first in Bern, Switzerland, and then in Berlin, Germany. In June 1941, Yamashita visited Europe again, where he met Adolf Hitler and Benito Mussolini as head of a military mission.

Because of his political commitment, he made an early enemy of Tojo Hideki, who would later become Army and Prime Minister. During the Second Japanese-Chinese War, Yamashita, as commander-in-chief of the Mixed Garrison Brigade and later as chief of staff of the North China Regional Army, pleaded repeatedly for an end to the conflict with China as quickly as possible so as not to jeopardize the good relations with Great Britain and the United States. Back in Japan, he was sent back into Manchurian exile after a short time with the Supreme War Council on the order of Army Minister Tojo Hideki.

In November 1941, Yamashita assumed command of the 2nd Army, which landed in Malaysia on December 8, 1941. Since his forces are only one-third of the Allied forces facing him, he decides to avoid grueling positional combat and advance through Malaysia to Singapore as quickly as possible. It is clear to him that he can only win the victory through a rapid advance that does not falter.

Yamashita achieves the almost impossible. On February 15, 1942, the main British naval base Singapore, which is considered impregnable, falls, and Yamashita and some 30,000 men of the Imperial Japanese Army capture 80,000 British, Indian and Australian soldiers. This military tour de force earned Yamashita the honorary title "Tiger of Malaysia". The British Prime Minister at the time, Winston Churchill, described the fall of Singapore as a disgrace and the greatest catastrophe and most serious capitulation in British military history.[124]

Situation meeting in the jungle of Malaysia. Lieutenant General Yamashita, in the foreground third from the right with his samurai sword, which accompanied him until his surrender. The sword was forged in the 17th century by Fujiwara Kanenaga.[125]

When Yamashita is to be granted an audience with Emperor Hirohito, Tojo Hideki, now Prime Minister, again orders his

---

124 https://en.wikipedia.org/wiki/Battle_of_Singapore
125 https://en.wikipedia.org/wiki/Japanese_invasion_of_Malaya

transfer to prevent this. Yamashita is appointed commander-in-chief of the 1st Regional Army, stationed on the Manchurian-Soviet border, with retroactive effect from July 1, 1942. In February 1943, he is promoted to general in this position. As the military situation for Japan continued to deteriorate, General Yamashita was recalled by the new government from his exile in China after the fall of Tojo Hideki's cabinet, and on September 26, 1944, he was appointed commander-in-chief of the 14th Regional Army stationed in the Philippines to organize defense against an impending Allied invasion.

General Yamashita had no more time for this, because only 10 days after his arrival in Manila, Allied troops landed on Leyte on October 20, 1944. With the advance of the Allies from early January 1945, General Yamashita ordered the evacuation of Manila and the retreat to better defended positions in the surrounding area. In the face of overwhelming power, General Yamashita Tomoyuki went to Allied headquarters on August 14 to negotiate the surrender of his troops. On September 2, 1945, he signed the surrender and handed over his samurai sword, which was forged by Fujiwara Kanenaga in the 17th century and had accompanied him on all his campaigns. The sword is now on display at the West Point Military Museum in New York.

During the invasion of Malaysia, Japanese soldiers attacked prisoners of war and civilians. To this day it is still disputed whether General Yamashita could be held responsible for these attacks at all because he was unable to prevent them. However, he had an officer responsible for the murder of patients and employees of a British military hospital and soldiers accused of looting executed and apologized to the survivors.[126] Furthermore, the Battle of Manila, for which Vice-Admiral Iwabuchi

---

[126]https://de.wikipedia.org/wiki/Schlacht_um_Singapur#Das_Massaker_im_Alexandra-Hospital

Sanji was responsible because he had defied General Yamashita's order to evacuate Manila, resulted in bloody fighting in which an estimated 100,000 Filipino civilians were killed in massacres by Japanese troops.[127]

As a result of this and other war atrocities committed by Japanese soldiers in Singapore and the Philippines, General Yamashita was brought before a U.S. military tribunal and charged as a war criminal. The trial began on October 29, 1945 and ended on December 7, 1945, with a verdict of guilty and the sentencing of General Yamashita to death by hanging. The case became known as the "Yamashita standard" and entered military jurisdiction as a precedent for the superior responsibility of officers when war crimes are committed under their high command, even if they did not order it or could not have prevented it. On February 23, 1946, the sentence against General Yamashita was carried out in the prison of Los Baños, Philippines. After he had climbed the gallows, he addressed those present with his last words:

"As I said in the Manila Supreme Court that I have done with my all capacity, so I don't shame in front of the gods for what I have done when I have died. But if you say to me "you do not have any ability to command the Japanese Army" I should say nothing for it, because it is my own nature. …When I have been investigated in Manila court I have had a good treatment, kindful attitude from your good natured officers who all the time protected me. I never forget for what they have done for me even if I had died. I don't blame my executioner. I'll pray the gods bless them. Please send my thankful word to Col. Clarke and Lt. Col. Feldhaus, Lt. Col. Hendrix, Maj. Guy, Capt. Sandburg, Capt. Reel, at Manila court, and Col. Arnard. I thank you."

---

[127] https://de.wikipedia.org/wiki/Schlacht_um_Manila_(1945)

"The process has been sharply criticized from the beginning and repeatedly since then. Many of the American officers involved in the prosecution and the conviction lacked experience in the front line and had previously received little, if any, training in law. In addition, Yamashita's defense counsel complained that they had been given too little time to prepare their client's defense and that requests for postponement of the trial had been rejected or ignored. The involvement of many Filipinos bent on revenge for the Japanese occupation and the effects of many years of inflammatory propaganda against Yamashita heated the atmosphere during the trial to such an extent that the court's award is not considered objective. ...In many cases the trial is also seen as a private feud of General Douglas MacArthur, who wanted to take revenge for the Japanese occupation of "his" Philippines. Moreover, MacArthur demanded that an example be made in this first war crimes trial of a Japanese to serve as a model for harsh treatment of other defendants in the upcoming Tokyo war crimes trials."[128]

For General Yamashita Tomoyuki, the path of the warrior was tragically fulfilled. The attitude with which he went to his death still deserves our greatest respect. Without fear of death and in the conviction not to be ashamed before the gods, he asks them to bless his executioner. The events of the war and the verdict by a U.S. military tribunal reveal the arbitrariness and double moral standards of the victor, who, despite considerable doubt about his guilt, pleads guilty to the defeated in order to set an example, while the victor, without need and to this day unatoned, killed about a quarter of a million people by dropping two atomic bombs on Hiroshima and Nagasaki, who either died directly in the dropping of the bombs or suffered and died cruelly from the late nuclear consequences.[129]

---

[128]https://de.wikipedia.org/wiki/Yamashita_Tomoyuki
[129]https://de.statista.com/statistik/daten/studie/1086264/umfrage/geschaetzte
-zivile-todesopfer-und-verletzte-in-hiroshima-und-nagasaki/

The statement that the bombing was done without necessity and were primarily decisions made for power-political considerations does not come from nowhere. On May 28, 1945, the then U.S. ambassador in Moscow had telegraphed to Harry S. Truman, who became the new President of the United States of America after the death of Franklin D. Roosevelt on April 12, 1945, that Soviet troops had taken up positions in Manchuria for the war against Japan. "Japan knew that it was lost. However, since Japan's government would not surrender unconditionally, Stalin had suggested accepting a Japanese peace offer and then enforcing its own objectives by jointly occupying and administering Japan."[130] Therefore, as early as 1965, the historian Gar Alperovitz questioned the U.S. government's claim that the atomic bombs dropped on Hiroshima and Nagasaki were intended to spare the lives of American soldiers. Rather, the drops had served the purpose of deterring the Soviet Union from advancing further in the Far East and demonstrating to it the power of the United States. [131] Other researchers further explain the dropping orders by saying that the use of the atomic bombs was to justify the high development costs of two billion dollars and their effectiveness should be tested on real targets. Racist motives are also mentioned, up to the portrayal of the missions as genocide.[132] [133] Thus, the use of the atomic bomb in Nagasaki in particular was, according to Martin Sherwin, "at best senseless, at worst murder of nations."[134]

---

[130]https://de.wikipedia.org/wiki/Atombombenabwürfe_auf_Hiroshima_und_ Nagasaki#Gegner_der_Abwürfe
[131]Alperovitz, Gar, Atomic Diplomacy: Hiroshima and Potsdam, Vintage Books 1965
[132]P. Joshua Hill, Professor Koshiro,Yukiko: Remembering the Atomic Bomb. FreshWriting 15 December 1997
[133]Knauß, Ferdinand, Ein Experiment mit 70.000 Toten. Zeit Online, 2009, https://www.zeit.de/online/2009/35/atombombe-hiroshima
[134]Scherer, Klaus, Interview mit Martin Sherwin, 3. August 2015,

General Dwight D. Eisenhower also later reported that he had advised President Truman against using the atomic bomb because the Japanese had already signaled a willingness to surrender and the United States should not be the first to use such weapons. But Truman wrote in his diary, "I believe the Japs will cave in before Russia intervenes." [135] This reckless scorched earth policy and Truman's apparent hatred of the Japanese is also reflected in paragraph 3 of the "Potsdam Declaration" of July 26, 1945, which states, "The full application of our military power, backed by our resolve, will mean the inevitable and complete destruction of the Japanese armed forces and just as inevitably the utter devastation of the Japanese homeland." Taking into account that Truman had received word one day before the conference began that the United States had successfully detonated an atomic bomb in the New Mexico desert, the wording in paragraph 3 of the Potsdam Declaration is a strong indication that Truman was already determined at that time to use the bomb. [136]

"Militarily, the bombs were really not necessary, and ethically they could not be justified anyway. This was confirmed after the war by many high-ranking generals who were themselves appalled by it. The most outspoken was Truman's own chief of staff, Admiral William Leahy, who accused the president of misleading him. Leahy had condemned the atomic bomb as a weapon of mass destruction. Truman had therefore specifically assured him that he would only use it to attack military targets.

In truth, however, America had then hit as many civilians as possible. Leahy was as outraged by this as his colleagues Ei-

---

https:// www.cicero.de/aussenpolitik/atombombe-auf-nagasaki-japan-haette-auch-ohne-bombe-kapituliert/59654
[135]https://de.wikipedia.org/wiki/Atombombenabwürfe_auf_Hiroshima_und_Nagasaki#Gegner_der_Abwürfe
[136]https://teachingamericanhistory.org/library/document/potsdam-declaration/

senhower, Nimitz, Spaatz and Arnold. Almost all the top military officers disagreed, we can go through the whole list. They all knew that Japan had long been defeated and wanted to surrender. How could our country be the first, they criticized, to still use that terrible bomb?

Indeed, a U.S. investigation after the war concluded that in official interviews, diaries, and other sources, both private and public, all senior military officers confirmed that the use of the bomb was not out of military necessity. Leahy's memoirs spoke explicitly of ethical standards common among barbarians of dark ages. He had not been taught as a soldier that this was how wars were fought. They could not be won by destroying women and children."[137] So far the US-American historian Martin Sherwin, according to Klaus Scherer probably the most proven expert on the biography of Robert Oppenheimer and the history of the atomic bombs.

---

[137] Scherer, Klaus, Interview mit Martin Sherwin, 3. August 2015, https:// www.cicero.de/aussenpolitik/atombombe-auf-nagasaki-japan-haette-auch-ohne-bombe-kapituliert/59654

August 14, 1945, Baguio, Philippines: General Yamashita Tomoyuki precedes his staff on their way to Allied headquarters to negotiate the surrender.[138]

[138] https://en.wikipedia.org/wiki/Tomoyuki_Yamashita

"The sword is the soul of the samurai. Whoever loses it is dishonored and subject to the severest punishment." The bitterest, if not the final moment in a samurai's life. The photo published in December 1945 shows Major General W.A. Crowther, British Commander of the 17th Indian Division, receiving in Thaton, north of Moulmein, Burma, the swords of the Supreme Commanders of the 3rd Japanese Army.[139]

---

[139] British Official Photograph, December 1945

---

[140] https://de.wikipedia.org/wiki/Kohima

神風
Kamikaze

"If one is confronted with two alternatives, life or death, one should choose death without hesitation." Aware of his inevitable fate, a pilot of the Shimpu Tokkotai boards the plane's cockpit with his samurai sword to his final flight.

In a book that deals with the meaning of samurai swords in the material battle, it is also necessary to remember those Japanese pilots who have made the term "kamikaze" world-famous since the Pacific War through their death-defying suicide attacks against allied naval units. For these pilots also took their swords on board their planes when they took off on their last mission. Due to the cramped cockpits, the doomed pilots also took shorter wakizashi or tanto on their last flight, in addition to katana or tachi. Thus the steely soul of the samurai accompanied Japan's knights of modernity on their way to death.

Much has been written about kamikaze missions and the men who flew them. Therefore we will only briefly discuss the origin and mission of the kamikaze units. Instead, the subject matter of this book makes it seem more purposeful to question the motives of those men who, with a clear mind, set off on their last flight in anticipation of their certain death. It is obvious that they did not correspond to the cliché of the blinded fanatics that Japanese pilots were often mistaken for. Rather, their motivations are reminiscent of the samurai's code of honor and sense of duty. The conviction that it was better to die than to be defeated and fall into disgrace was deeply rooted in the mindset of the Japanese. Honor, personal and family honor, traditionally had the highest value in Japan. This also led to the reporting of volunteers. But also the pressure of group pressure and the resulting high expectations, which still determine the actions within the Japanese society today, may have driven some of the mostly young pilots to their decision.

First, however, the historical origin of the term "kamikaze" will be briefly discussed, a term that is increasingly being forgotten in the western world due to its predominant use in connection with suicide attacks.

Having previously subjugated China and Korea, the Mongols tried to conquer Japan twice towards the end of the thirteenth century. The first time they landed in 1274 with an estimated force of 900 ships and 30,000 men in the Bay of Hakata on the island of Kyushu in southern Japan. The far inferior samurai desperately resisted the overwhelming superiority, until a violent storm destroyed the invasion fleet. In 1281 the Mongols returned once again. With an armada of 4500 ships and an estimated 140,000 men they attacked Japan again. But this time the samurai were prepared for the attack. With small, agile boats they boarded the ships of their opponents, set them on fire and killed the crews in a man-to-man fight. Inland, a gigantic stone wall 20 kilometers long in the bay of Hakata prevented the attackers from advancing any further. After two months of fierce fighting, a typhoon sent the invasion fleet to the bottom of the sea again. The Mongols, who had lost about 100,000 men, experienced the greatest defeat in their history. Believing that both times the gods had summoned the storm and had come to their aid, the Japanese called the storms "Kamikaze" ("Divine Wind"). But back to the Japanese fighter pilots of the Second World War.

神風特攻隊
Shimpu Tokkotai

The pilots who flew the Kamikaze missions belonged to the *Shimpu Tokkotai*. The *Shimpu Tokkotai* were air combat units of the Imperial Navy, which were recruited from naval pilots shortly before the end of World War II and went down in history for their dreaded attacks against allied naval units. *Tokkotai* was the military abbreviation for *Tokubetsu Kogekitai ("Special Attack Force")*. The founder of this force was *Admiral Onishi Takijiro*, who took command of the 1st Imperial Japanese Air Fleet on Luzon in October 1944.

Three months earlier, from June 19 to 20, 1944, the largest naval aircraft carrier battle of the Pacific War had raged near the Marianas, in which the Japanese suffered heavy losses. The outcome of the battle was due to the enormous economic efforts of the United States, which had led to the rapid armament of the American Pacific fleet and the use of improved weapons systems. The numerical superiority of the U.S. Navy, the use of new armored fighter aircraft and the equipping of the ships with an unprecedented number of anti-aircraft guns significantly determined the outcome of the battle, which ended for Japan with three sunken aircraft carriers and more than 400 destroyed aircrafts.

The reasons for the enormous losses of the Japanese Air Force were on the one hand the lack of experience of the young pilots and on the other hand air combat tactics newly developed by the Americans. The Americans had found out that the Japanese engines had catastrophic performance gaps at certain altitudes and speeds, due to research into the flight performance of captured Japanese aircraft. At altitudes above 2000 m, the American aircraft were clearly superior to the Japanese.

Based on these findings, the Americans led the Japanese pilots to attack at higher altitudes. The young Japanese pilots were too inexperienced to understand this tactic of the Americans. Instead of going into a position at low altitude where their planes would have been superior to the American ones in maneuverability, they followed the Americans into high altitudes and were involved in hopeless air battles. This air battle later went down in war history as "The Marianas Turkey Shoot".[141]

Admiral Onishi Takijiro, founder of the Shimpu Tokkotai.

---

[141] Barrett Tillman, Carrier Battle in the Philippine Sea: The Marianas Turkey Shoot, Specialty Press, 1994

After this defeat, Onishi Takijiro no longer had enough fighter planes at its disposal to decisively hit the US carrier fleet with powerful bombing raids. On the advice of the Tenno he is said to have decided to sacrifice the few remaining aircraft crews in suicide attacks.[142] It is more likely, however, that Emperor Hirohito was unaware of Onishi's plans to establish the *Shimpu Tokkotai*. For the Tenno, as a direct descendant of the sun goddess *Amaterasu*, was more of a spiritual leader of Japan. Participating in daily politics or issuing orders was neither within his competence nor was it expected of him. The power lay exclusively with the military government. When Emperor Hirohito was told of the first successful suicide attack, he is said to have welcomed the success but regretted the fate of the pilot.[143]

In addition to military considerations, propaganda considerations also played a role in the establishment of the *Shimpu Tokkotai*. The voluntary sacrificial death of the pilots was supposed to serve as an incentive for further recruits to follow the fallen heroes on their glorious path. The Navy only accepted volunteers for the *Shimpu Tokkotai*. "Married, first-born and only sons were rejected.[144] This led in the case of *First Lieutenant at Sea Hajime Fujii (Kaigun-Chui Hajime Fujii)* to his wife drowning herself along with their two children so that he could accompany his students to their deaths.

Hajime Fujii was born on Aug. 30, 1915, in Joso, Ibaraki Prefecture, Japan. As a young man, he decided to pursue a career in the army and participated in the Second Japanese-Chinese War (July 7, 1937 - Sept. 9, 1945). After being wounded, he met his future wife Fukuko in the field hospital, where she worked as a nurse. The marriage produced two delightful daughters – Kazuko and Chieko – to the great joy of Hajime

---

[142] https://de.wikipedia.org/wiki/Ōnishi_Takijirō
[143] https://de.wikipedia.org/wiki/Shimpū_Tokkōtai
[144] https://de.wikipedia.org/wiki/Shimpū_Tokkōtai

and Fukoku. Due to his wounding, Hajime Fujii did not return to the front, but was transferred to the Army Air Corps Academy, where he graduated in 1943. Since his hand wound did not allow him to fly himself, he trained young pilots in character and mental strength as a company commander at the Kumagaya Army Aviation School in Kaitama.

When his students were also selected for kamikaze missions, he decided to accompany them on their last flight because he did not want them to go to their deaths alone. He made a request to that effect, which was denied. The decision to accompany his students was by no means the result of fanatical nationalistic delusion, but of a well-considered and serious decision, which must be regarded with the greatest respect as probably the most consistent form of exercising responsibility that a superior can assume for men entrusted to him. For when more and more of his students were sent off on "Tokko" missions and did not return, Hajime Fujii was tormented by the thought that he was betraying the young pilots entrusted to him. While indoctrinating them ideologically and invoking loyalty and patriotism in them, he stayed behind while his students met their deaths. "He felt like a hypocrite, which is why he appealed again to the army to let him die."[145] This second request was also denied. His wife Fukoku understood that her husband would suffer forever from the personally felt betrayal of his students and lose his honor, thus falling into disgrace. And Fukoku and his children would be the cause of his loss of honor and disgrace because, as a married man, he was ineligible for "Tokko" missions.

"On the morning of December 14, 1944, while her husband was in Kumagaya, Fukuko dressed in her finest kimono. She did the same with three-year-old Kazuko and one-year-old

---

[145] https://www.warhistoryonline.com/world-war-ii/the-tragic-tale-of-hajime-fujii-a-kamikaze-fighter.html

Chieko. Finally, she wrote a letter to her husband, urging him to do his duty to the country and not to worry about his family. They would be waiting for him. Then she wrapped Chieko in a cloth backpack and strapped the baby to her back. She took Kazuko by the hand and walked to the Arakawa River near the school where her husband taught. She took a rope, tied Kazuko's wrist to her own and jumped into the freezing water. Police found the bodies later that morning and Hajime Fujii was taken to the place where they lay laid out."[146]

The following evening, Hajime Fujii wrote a letter to his wife Fukuko and his daughters:

"It was a gusty and cold day in December.

You lost your lives in Arakawa River. You died with your mother before I die in order to meet my determination that I give my life for the country. I mourn my young children's death. I hold you dear, who died as if you were smiling.

I shall get after you soon. When we meet next time, let me held you in my arms without hesitation, and let's go beddy-bye on my chest.

Please wait for me without crying until I go to you.

Dear Kazuko-chan, please take care of Chieko-chan when she cries.

Then, Good-bye for a while.

I'm sure to serve with great distinction in the front, and I will go to you.

Well, Kazuko-chan and Chieko-chan, please wait for me until then."[147]

---

[146] https://www.warhistoryonline.com/world-war-ii/the-tragic-tale-of-hajime-fujii-a-kamikaze-fighter.html
[147] http://www.chiran-tokkou.jp/learn/pilots/FujiiHajime.html

Then he cut off his little finger and wrote his third request to the army with his blood. The letter to his wife and children and his last will and testament are preserved at the Chiran Peace Museum in Minamikyushu-shi, Kagoshima Prefecture, Japan.

On February 8, 1945, Hajime Fujii was appointed commander of the 45th Shinbu Squadron, which he christened Kaishin ("Joyful Spirit"). Just before dawn on May 28, 1945, the nine *Kawasaki Ki-45* aircrafts, each with a pilot and gunner aboard, set out for Okinawa. North of Okinawa, 50 miles offshore, they encountered an American naval task force escorted by the destroyer "USS Drexler" along with other destroyers. Dr. Rex Davis, who was serving as quartermaster on one of the escort ships at the time, later reported:

"Suddenly the *Drexler* took a hit. A twin-engine bomber had been shot up but made it through. Veering wildly and burning, its suicide run was almost a stumbling run into the *Drexler* and paralyzed her. ... She was now being dived upon by another twin-engined bomber that was being chased by two *Corsairs*. As the bomber was about to consummate its dive into the *Drexler*, it missed the mark. It overflew and was so low that it looked as if it were doomed to splash into the sea. But it did not as it barely skimmed the water. Then it rose and turned to the left. The *Corsairs* chasing it were going too fast to follow the turning bomber. The Japanese pilot maneuvered skillfully, and it was evident to us who had seen many Kamikaze dives that this was no new untrained pilot. He flew his plane into a vulnerable area, aft of amidships in front of the torpedo tubes. All exploded as the suiciding aircraft hit the *Drexler*, sending part of the opened-up ship into the sky, while most of the ship parted and opened up to the seas. Initially the *Drexler* was obscured by a cloud of smoke. I had noted the time when hit, and when again visible, the bow was up and going down fast. In 49 seconds from being hit by the second Kamikaze, she was gone."[148]

First Lieutenant Hajime Fujii was a gunner in one of the aircraft. He was posthumously promoted to major.[149] Let us wish him to be with his family.

First Lieutenant at Sea Hajime Fujii

---

[148] http://www.kamikazeimages.net/stories/lcsl1114/index.htm
[149] http://www.chiran-tokkou.jp/learn/pilots/FujiiHajime.html

As we know today, even the pilot *Seki Yukio*, who later was stylized from Japanese propaganda to a godlike *kami* and led the first successful kamikaze mission of World War II, was no fanatical hotspur. Born on August 29, 1921, in *Iyo Saijo*, a small town in *Shikoku Prefecture*, *Seki* was already enthusiastic as a student for the Navy. At the age of 17, he applied to the Imperial Naval Academy and after successful completion of his studies, he began his service on the battleship *Fuso*. After his promotion to lieutenant he is transferred to the aircraft carrier *Chitose*. In the course of the war Seki fought in various battles, including the *Battle of Midway*. In 1942 he returned to Japan and was accepted into the Naval Aviation Academy. After training as a carrier-operating dive bomber pilot, *Seki* was assigned as a flying instructor in January 1944 due to his talent and outstanding skills.

In October 1944, *Seki* became the leader of the 301st Battle Squadron and was transferred to the 201st Naval Aviation Wing on the Philippines. There *Group Commander Tamai Asaichi* summoned him to report on October 19. When *Seki* reports, he sees that *Captain Inoguchi Rikihei, Admiral Onishi's* chief staff officer, is also present. It is revealed to him that *Admiral Onishi* is planning a suicide attack on an American naval unit and is considering entrusting *Lieutenant Seki* with this task. When *Seki* is asked by *Tamai* whether he agrees with this, it is immediately clear to him that this is not a question but an order. His code of honor as a Japanese officer leaves him no choice. After a short silence he answers: "Please put me in charge of this task." And although *Seki* has only recently married and thus does not meet the selection criteria for accepting volunteers for kamikaze missions, his successful career in the Navy and as a naval aviator makes him the ideal showcase candidate for this mission for the officers present.

On the morning of October 20, Admiral Onishi Takijiro personally performs the farewell ceremony for the small combat squadron under the command of *Lieutenant Seki Yukio*. Witnesses later reported that *Admiral Onishi* had tears in his eyes when he shook the pilots' hands one last time. After a last meeting with *Group Commander Tamai* and *Captain Inoguchi, Seki* retired to his quarters to write to his wife Mariko and his parents. In the tradition of the samurai, he writes a poem for the flight students under his command as a farewell:[150]

"Fall my Pupils,
My cherry blossoms,
Just as I will fall
In service to our country."

In view of his inevitable fate, the poem reveals the young squadron leader's identification with the self-image of the Samurai. For just as the cherry blossoms fall from the tree in their most beautiful blossom, so it was considered desirable for the Samurai to die in the blossom of their lives in battle according to the principles of honor and loyalty.[151] The Japanese proverb *"The most beautiful of all flowers is the cherry. The noblest among men is the samurai."* gives an idea of the emotions and steadfast loyalty his poem must have aroused in the young pilots who flew with him, as it was tantamount to a knighthood.

In the letter to his wife Mariko, he apologizes that he was destined to fall on his mission and that he could not do more for her. But he knows that as the wife of a soldier she is prepared for such a moment. The letter ends with the words: "Please take good care of your parents. Now that I am saying goodbye,

---

[150] Emiko Ohnuki-Tierney, Kamikaze, Cherry Blossoms and Nationalisms, 2002
[151] http://www.bushido.de/philsophie.htm

countless shared memories are awakening in me. All the best for little Emi *(Mariko's younger sister)*. Yukio"[152]

Also from the letter to his parents it does not emerge with any word that he was as it were assigned to the mission and how he really thinks about it. He limits himself merely to the suggestion that he sees Japan on the brink of defeat and explains that anyone who has embarked on a military career has no choice in such a situation and must pay his debt to the emperor. The letter ends with the words: "I surrender to my fate. I greet you all and look forward to the end with composure. Yukio"[153]

However, before his last flight, Lieutenant Seki reveals his personal inner conviction about the suicide mission to a reporter: "Japan is at the end of its rope when it is forced to kill one of its best pilots. I am not going on this mission for the Emperor or for the Japanese Empire. I am going because I was ordered to!"[154] In fact, Lieutenant Seki was convinced that as a talented and experienced pilot on many enemy flights, he could have served his country far better than with this one mission alone.[155]

---

[152] Emiko Ohnuki-Tierney, Kamikaze, Cherry Blossoms and Nationalisms, 2002

[153] Emiko Ohnuki-Tierney, Kamikaze, Cherry Blossoms and Nationalisms, 2002

[154] http://en.wikipedia.org/wiki/Yukio_Seki

[155] https://ww2db.com/person_bio.php?person_id=297

Seki Yukio as a graduate of the Imperial Navy Academy in 1939.

On October 25, 1944, the *"Shikishima Squadron"* consisting of five Mitsubishi A6M2 Zero dive bombers under the command of Lieutenant Seki Yukio and accompanied by four escort fighters took off from *Mabalacat Air Base (Manila, Philippines)* for its last mission. At 10:47 am, the squadron sighted its targets in the sea area of the *Gulf of Leyte*. While the escort fighters are involved in fierce air battles with American fighter planes, Lieutenant Seki and his men carry out the first successful Kamikaze attack in the history of World War II. In close cooperation with their squadron leader, the young pilots nosedive their planes in cold blood towards their targets. Although the pilots who fly with Lieutenant Seki are relatively inexperienced, four of the five Zeros, all of which are equipped with 250-kilo bombs, hit their targets. During the attack, all the aircraft carriers are damaged except for *Fanshaw Bay*.

At 10:51 am Seki's fate is fulfilled when his Zero shatters on the flight deck of the *"USS St. Lo"*. Observers later describe that a Zero broke away from the formation and nosedived onto the carrier at an altitude of about 100 feet. Despite the massive barrage of fire, the pilot managed to break through the deadly curtain of detonating flak grenades. It is further reported that the pilot appeared completely calm and level-headed as he kept his plane on its deadly course. The pilot safely releases his bomb, which hits the flight deck in the middle. Seconds later, the aircraft also hits the deck. The bomb, weighing 250 kilos, has penetrated the flight deck and explodes on the hangar deck, where aircrafts are being refueled and equipped with ammunition. Suddenly the kerosene catches fire and triggers numerous explosions, among others in the torpedo and bomb magazine. Wrapped in a huge sea of flames, the "USS St. Lo" sinks within 30 minutes.

A single Zero caused the sinking of an aircraft carrier. Or, as one observer later put it: "A pilot, a Zero, a bomb, an aircraft carrier." The cold-bloodedness and skill the pilot displayed in the execution of his deadly mission allowed only one conclusion: the pilot was none other than Lieutenant Seki Yukio himself. It was the first successful Kamikaze attack of the Second World War in which an aircraft carrier was sunk. Of the 889 men on board, 113 were killed or reported missing, 30 others later succumbed to their injuries. Also for Lieutenant Seki Yukio and his comrades this attack was a flight to certain death.

The last Kamikaze mission of the Second World War is commanded by *Admiral Ugaki Matome*, and he himself flies with them. Born in *Okayama, Okayama Prefecture* in 1890, *Ugaki* joined the navy at the age of 16. After his basic training and service on various ships, he successfully completed the Naval Academy, later the Naval Staff College. Due to his popularity with superiors and men, *Ugaki's* career is on the rise. From 1928 to 1930 he represented Japan as a naval attaché in Germany. In 1935, *Ugaki* was assigned to the Japanese Combined Fleet as a staff officer for one year before he took command of a cruiser and then a battleship. In 1938 he was promoted to rear admiral.

Admiral Ugaki Matome with his samurai sword in Kai-Gunto mounting. The brown portepee, which is uniformly unicolored (brown) for naval officers of all ranks, is clearly visible.

During World War II, *Ugaki* served as commander of the Imperial Japanese Navy Battle Fleet and commanded the 1st Battleship Division with the two super battleships *"Yamato"* and *"Musashi"*.[156]

The assembled fleet staff of the 1st battleship division on the "Yamato". Admiral Ugaki Matome, front row, fifth from left. Next to him and sixth from the left, Yamamoto Isoroku, Chief of Staff and Commander-in-Chief of the United Japanese Fleet. The "Super Battleship" Yamato was the first ship of the Yamato class in World War II. At 46 centimeters, the heavy artillery had the largest caliber ever used on battleships. The ship was built from 1937 to 1941 at the naval shipyard in Kure, Japan, and was in service together with its sister ship "Musashi" during the Pacific War. The Yamato was sunk by US carrier aircraft on April 7, 1945, some 300 kilometers south of the Japanese island of Kyushu.[157]

---

[156] http://de.wikipedia.org/wiki/Ugaki_Matome
[157] https://de.wikipedia.org/wiki/Yamato_(Schiff,_1941)

In February 1945, *Ugaki* was appointed commander of the 5th Navy Air Force Air Fleet in *Kyushu*. In this capacity, *Ugaki* further developed the ideas of *Onishi Takijiro* and planned large-scale kamikaze operations. Due to the now clear superiority of the American Pacific Forces, he sees this as a last resort to avert a possible invasion and military defeat by Japan. As dive bombers, *Ugaki* also uses outdated *Mitsubishi A6M3 Rei-sen/Zero-sen*, which are no longer suitable as fighter planes, but are still available in large numbers. In the firm belief that he would be able to turn the fortunes of war around once again, he ordered the Kamikaze deployment of several hundred aircraft in the Battle of Okinawa (April 1 to June 30, 1945) against the US fleet.

On August 15, 1945, the Tenno announced Japan's surrender on the radio with the "Imperial Decree on the End of the War" and at the same time called on the armed forces to lay down their arms. When *Ugaki* hears the speech, he writes in his diary that he has received no "official" order to cease fighting. On the basis of later reports from his officers and soldiers, it is considered certain that *Admiral Ugaki* intended to die with honor in the event of Japan's defeat in order to "display the true spirit of a Japanese warrior".[158]

Immediately after the proclamation of Emperor Hirohito, *Ugaki* announces that he is looking for volunteers who are willing to fly with him in a final kamikaze attack against the U.S. fleet. Thereupon the crews of eleven dive bombers of the 701st Group report. Actually *Ugaki* only wanted to fly with five planes, but the entire unit decides to accompany its commander. Before he boards one of the planes, he has the rank insignia removed from his dark green navy uniform. Then he has his picture taken in front of one of the planes. Since he is not an aviator himself, he climbs into the rear section of a *Yokosuka*

---

[158] http://de.wikipedia.org/wiki/Ugaki_Matome

D4Y3, which he has used several times for connecting flights, with the second crew member. On his flight to his death, *Ugaki* takes the *wakizashi* with him, which *Admiral Yamamoto Isoroku* had once presented him with as a personal gift.[159]

The attack formation takes off at sunset and reaches *Iheyajima Island* near *Okinawa* at 19:40 local time. It is not known for sure how many of the eleven planes hit or damaged a ship. Of the eleven planes that took off, however, three returned to the base, with crews explaining their return in accordance with engine problems. In fact, however, it can be assumed that these three planes returned on *Ugaki's* order to testify about the mission.

The next morning the Americans find the smoking debris of a Japanese *D4Y3* and the remains of the pilot's cockpit with three dead. All other aircraft had only two crew members on board, as is usual with this type of aircraft. One of the three dead is wearing a dark green navy uniform. Next to the leech, the Americans find the samurai sword that *Admiral Ugaki* took with him on the last Kamikaze mission of World War II. The sword was given to *General Douglas McArthur*. He later donated it to the *U.S. Merchant Marine Cadet Corps Academy* in *Kings Point, New York State*, where it is still kept today.

Admiral Ugaki Matome went down in history as *"The Last Kamikaze"* in this questionable chapter of Japanese warfare. His deployment on the last day of the war was criticized as pointless even within the Japanese Navy, especially since he took more men with him to their deaths. On the other hand, it was recognized that he, as an organizer of large-scale kamikaze operations, in which many army and even more naval pilots died, had followed these men to their deaths consistently and honorably in the samurai tradition. His diary, as an important contemporary document of the Pacific War, fell into the hands

---

[159] http://de.wikipedia.org/wiki/Ugaki_Matome

of the Americans against his expressed will and remained preserved for posterity. It was translated into English and has been published as a book.[160]

"The Last Kamikaze": Last day of the war, August 15, 1945, Admiral Ugaki Matome is photographed shortly before the start of the last Kamikaze mission of World War II in front of the Yokosuka D4Y3, which he will climb shortly after. The rank insignia on his uniform are already removed. In the cockpit, Ugaki takes a wakizashi that Admiral Yamamoto Isoroku once presented to him as a personal gift.

The decisive turnaround could not be brought about by the pilots of the *Shimpu Tokkotai* through their sacrificial death.

---

[160] Fading Victory: The Diary of Admiral Matome Ugaki, 1941-1945, University of Pittsburgh Press, 1991

Nevertheless, they inflicted severe losses on the Allies with their death-defying attacks. There are no exact figures, but it is estimated that at least 47 Allied ships were sunk by kamikaze attacks, from the torpedo boat to the aircraft carrier. Another 300 ships were damaged, some of them severely. Apart from the actual number of damages or sinking, another consequence of the Kamikaze attacks cannot be ignored: It was the nervous strain on the crews of the ships that finally brought the number of war neuroses on the American side to a level that gave the naval leadership serious cause for concern.[161]

By the end of the war, over 4,000 Japanese pilots had lost their lives in Kamikaze missions. Most of them were shot down by enemy fighters or failed in the massive defensive fire of the naval flak before they could fulfill their mission. Only about 14% reached their targets. Most of these men were soldiers who were aware of the hopelessness of their mission and reported to the missions out of a deep sense of honor or duty. However, there were also aviators who were simply ordered to go on the mission or such who were not convinced of their mission, but who submitted to group pressure. Sacrificial death was proclaimed as a heroic deed and was considered a war duty for the selected, if victory could be achieved.

In return, the propaganda conjured up the *spirit of bushido*. Unbreakable loyalty and the willingness to die at any time in the bloom of their youth without hesitation in faithful fulfillment of duty determined the samurai's ideal of life. This willingness to make sacrifices and the Buddhist basic attitude that death is a welcome conclusion to life on earth are also reflected in a poem written by Admiral Onishi Takijiro in the tradition of the samurai.

---

[161] https://de.wikipedia.org/wiki/Shimpū_Tokkōtai

Today in flower,
Tomorrow scattered by the wind —
Such is our blossom life.
How can we think its fragrance lasts forever?[162]

Onishi Takijiro

On August 16, 1945, just one day after the capitulation of Japan, the founder of *Shimpu Tokkotai, Admiral Onishi Takijiro*, committed seppuku. In a farewell letter he apologized to the pilots he had sent to their deaths and asked the fallen pilots and their families to accept his suicide as a personal penance. Kodama Yoshio, who accompanied him on his death, later reported that *Admiral Onishi Takijiro* renounced a second *(kaishakunin)* who could have shortened his death struggle by beheading. Due to the injuries he had inflicted on himself, his agony lasted over 15 hours.[163]

The sword used by *Admiral Onishi Takijiro* in seppuku is now kept in the *Yushukan Museum* in *Yasukuni Shrine*. Its ashes were divided between two resting places. One half rests in *Soji-ji (main shrine)* in *Tsurumi, Yokohama, Kanagawa Prefecture*, the other half in the municipal cemetery in the former *Ashida* in *Hyogo Prefecture*.[164]

With the end of this chapter of Japanese war history and the sacrificial attitude of Japanese soldiers during World War II, the circle is closed around the importance of the samurai swords in the battle of material. Even though they had lost their practical importance as weapons on the ground, especially in the cockpits of pilots or on board ships, they had become an important link between the samurai and Japan's warriors of the twentieth century. Mori Masahiro has already anticipated the

---

[162] https://www.poetryfoundation.org/poets/takijiro-onishi
[163] https://de.wikipedia.org/wiki/ Ōnishi_Takijirō
[164] https://en.wikipedia.org/wiki/Takijirō_Ōnishi

answer in his article "SOLDIERS AND GUNTO", quoted at the beginning of this article, by stating that the Japanese sword mentally brought together the Spirit of Bushido and the Imperial soldiers. Forged in the Spirit of Bushido, imbued with spiritual power, worshipped for centuries as the soul of the Samurai and as a status symbol of a warrior elite, the Samurai swords reminded their wearers of the virtues of the Samurai in the inferno of material battles and made the soldiers of the Emperor rise above themselves in the face of death.

"The way of the warrior is to die."
Missions of the Shimpu Tokkotai

Picture Section

Pilots of the Shimpu Tokkotai proudly pose with their samurai swords before their last flight. When the pilots took off for enemy flight, they took their swords on board. The photo shows the former Kamikaze pilot Toshio Yoshitake, in the picture on the right, with his comrades Tetsuya Ueno, Koshiro Hayashi, Naoki Okagami and Takao Oi (from left to right). The picture was taken on November 8, 1944, at the Choshi Army Airfield, east of Tokyo, just before takeoff. All pilots who took off that day with Toshio Yoshitake were killed. Yoshitake was the only one to survive because he was shot down by an American fighter plane and rescued by Japanese soldiers after a crash landing.

The group photo shows all comrades who started with Toshio Yoshitake on November 8, 1944. Such photos were taken less for the family album, but served primarily for propaganda purposes. The defiance of death that the young pilots radiated and their heroic sacrificial deaths were intended as an incentive for other recruits to follow the fallen heroes on their glorious path. The smiles of the young men do not betray how they really thought about their mission.

October 25, 1944, Mabalacat Air Base, Manila, Philippines: Admiral Onishi Takijiro (in the foreground) saying goodbye to "Shikishima Squadron" under the command of Lieutenant Seki Yukio (in the background second from the left, saluting). Admiral Onishi is said to have had tears in his eyes when he shook the pilots' hands one last time. After take-off, the squadron sighted its targets at 10:47 am in the sea area of the Gulf of Leyte. Four minutes later, at 10:51 am, the fate of the squadron leader Lt. Seki Yukio is fulfilled when his Zero crashes on the flight deck of the "USS St. Lo" as ordered.[165]

---

[165] https://de.wikipedia.org/wiki/Onishi_Takijiro

"A pilot, a zero, a bomb, an aircraft carrier": On October 25, 1944, Lt. Seki Yukio sunk the "USS St. Lo" in a suicide attack. Despite the massive barrage of fire, Lieutenant Seki Yukio succeeds in breaking through the deadly curtain of exploding flak grenades. Then he takes a decisive dive. The experienced dive bomber pilot releases his bomb unerringly, before he crashes on the flight deck with his Zero shortly afterwards. The 250-kilo bomb has penetrated the flight deck and explodes on the hangar deck, where aircrafts are being refueled. Suddenly the kerosene catches fire and triggers numerous explosions, including in the carrier's torpedo and bomb magazine. Wrapped in an enormous sea of flames, the St. Lo sinks within 30 minutes. It was the first successful attack by the Shimpu Tokkotai. For Lieutenant Seki Yukio it was a flight to certain death.

October 30, 1944, Philippine Sea: During the operation against Japanese units on Luzon, Philippines, the aircraft carrier "USS Belleau Wood" is hit in a kamikaze attack. While part of the flight deck crew fights the flames, other try to bring undamaged torpedo aircrafts of the type "Grumman TBM Avenger" to safety from the flames. In the background, the burning "USS Franklin" which received two Kamikaze hits in the attack. After the end of the Second World War, the "USS Belleau Wood" was initially placed under the command of the US reserve fleet. Later, the carrier then served as the "Bois Belleau" in the French Navy from 1953 to 1960, taking part in the war in Indochina and Algeria.[166]

---

[166] https://en.wikipedia.org/wiki/USS_Belleau_Wood_(CVL-24)

November 25, 1944, 12:55 pm local time, Philippine Sea: On the battleship New Jersey, soldiers must watch inactively from their command post as a Shimpu Tokkotai pilot completes his deadly mission. The photo shows the Mitsubishi A6M Zero just before it collides with the flight deck of the aircraft carrier "USS Intrepid".

Due to the force of the explosion and leaking kerosene, the flight deck of the "USS Intrepid" is immediately engulfed in bright flames that rapidly spread.

November 25, 1944, 12:52 pm local time, Philippine Sea: The "USS Entrepid" is reported to have two approaching Japanese *Zeros*, which approach the aircraft carrier from astern at an altitude of about 1.2 km and a distance of about 13 km. "Since there were many of their own aircrafts in the air, the gunners on board had to watch the aircrafts closely until they opened fire. So it was not until a minute later that a battery of the carrier began to fire at the left fighter. It exploded just above the water at a distance of 1.3 km astern. Since a *Hellcat* and an *Avenger* were operating in the same direction, the captain ordered the fire to cease. The starboard gunners, however, continued firing and were able to shoot down another low-flying Japanese aircraft. The second of the originally sighted fighters approached the stern in low-flying flight, evaded a *Hellcat* and dove away through the hail of bullets from the still incessantly firing rear batteries. Then the *Zero* went into a stalled climb at a distance of one kilometer, which brought it to a height of about 150 meters, dropped over one wing and crashed onto the flight deck of the Intrepid at 12:55 pm. The bomb penetrated the too weakly armored flight deck and exploded in the pilot's lounge, which was unoccupied at that time. Nevertheless, 32 men of the carrier were killed on deck and in the immediate vicinity of the bomb's explosion site.

...Only three minutes after the impact, the men in the gun emplacements spotted two more *Zeros* flying at 100 meters above the water. The dense smoke from the burning deck moving in exactly this direction had largely obstructed the starboard gunners' view. The port gunners managed to shoot down the first machine at a distance of 1.3 km, but the second one dived under the shells with rapid right-left turns. The pilot then pulled the plane up, tipped over one wing and crashed onto the flight deck at 12:59 pm. The bomb he was carrying exploded on the hangar deck."

The report bears witness to the cold-bloodedness and flying skills of the young Japanese pilots, but also to the determination with which they fulfilled their mission, even though they knew they were flying to certain death. [167]

---

[167] https://de.wikipedia.org/wiki/USS_Intrepid_(CV-11)

January 5, 1945, Gulf of Lingayen. In the course of landing operations in the Gulf of Lingayen, the "USS Louisville" receives two kamikaze hits causing medium damage. The losses are 32 dead and 56 injured. The ship itself remains operational and continues its mission for the time being. The "USS Louisville" was a heavy cruiser of the Northampton class. She was put into service on January 15, 1931 and survived the Second World War despite numerous missions. For her long and successful service in the Pacific War, the "USS Louisville" was awarded a total of 13 "Battle Stars".[168]

[168] https://de.wikipedia.org/wiki/USS_Louisville_(CA-28)

April 11, 1945, Battle of Okinawa (April 1 to June 30, 1945): "The way of the samurai is death." In a breakneck approach, a Japanese Kamikaze pilot flies under the raging fire of the ship's flak just above the waves and heads directly for the "USS Missouri" with his Mitsubishi A6M Zero. The Zero collided with the armor of the USS Missouri below the main deck, causing only minor damage and no casualties. The "USS Missouri" was put into service on June 11, 1944, and also took part in the Battle of Iwo Jima during the Pacific War. On September 2, 1945, the Japanese surrender was signed aboard the "USS Missouri".[169]

---

[169] https://de.wikipedia.org/wiki/USS_Missouri_(BB-63)

May 11, 1945, Battle of Okinawa: The "USS Bunker Hill" is hit by two kamikaze planes within 30 seconds. 372 soldiers die, 264 are wounded. The aircraft carrier was so badly damaged that it had to return immediately to the naval shipyard in Bremerton, Washington State. The "USS Bunker Hill" was an Essex-class aircraft carrier and was put into service on May 24, 1943. During the Pacific War, the carrier took part in numerous battles, including the Battle of the Philippines, the Battle of Iwo Jima and the attack on the Japanese fleet in the East China Sea. Four Japanese destroyers, a cruiser and the super battleship "Yamato" were sunk. In November 1966 the "USS Bunker Hill" was decommissioned and in May 1973 it was sold for scrapping.[170]

---

[170] https://de.wikipedia.org/wiki/USS_Bunker_Hill_(CV-17)

"Today in flower, tomorrow scattered by the wind — such is our blossom life. How can we think its fragrance lasts forever?" In anticipation of their inevitable fate, the young pilots seem to mentally merge with their swords. Infused with spiritual power, revered for centuries as the soul of the Samurai and as a status symbol of a warrior elite, the swords reminded their wearers of the virtues of the Samurai and made them rise above themselves at the hour of their death.

Property dies,
Clans die,
You yourself die like them;
I know one thing,
That lives forever:
The dead men's fame of deeds.[171]

Glossary

This glossary is not a complete list of the Japanese sword terminology. It should serve the interested reader only for a better understanding, as far as the description of the swords or the translation faithfulness made the use of the Japanese terms seem reasonable. Further explanations of terms were included, as far as the subject matter made this seem reasonable.

Amaterasu-o-mi-kami	Most important Shinto deity; personification of the sun and the light, founder of the Japanese imperial house.
ashi	Literally "foot"; narrow stripes of *nie* or *nioi*, running vertically from the tempering line *(hamon)* to the cutting edge.
Avenger	The Grumman TBF Avenger was the standard torpedo bomber of the US Navy at the end of World War II.
bakufu	Military government of the warrior nobility under the leadership of a *shogun*.
bizen-den	Bizen school
bizen-zori	Sword blade with the curvature near the tang, also known as *koshi-zori*.
bo-hi	wide fuller
bo-hi – hisaki-agari	*bo-hi* running beyond the *yokote*

bo-hi – kaki-nagashi	*bo-hi* running into the tang
bo-hi – kaku-dome	*bo-hi* with square end
bo-hi – maru-dome	*bo-hi* with round end
boshi	tempering line in the tip of the sword
boshi – kaen	flame shaped *boshi*
boshi – kaeri	turn back of the *boshi* to the *mune*
boshi – ko-maru	small round *boshi*
boshi – o-maru	large round *boshi*
Boshin-Krieg	War (1868-1869) between the Tokugawa-Bakufu and the imperial troops It ended with the defeat of the *bakufu* and sealed the fate of the samurai.
bu	Japanese scale unit; 1 *bu* = 0,303 cm.
buke	warrior nobility, Samurai
bushido	Literally "way *(do)* of the warrior *(bushi)"*. Code of conduct and honor of the samurai in the late Japanese Middle Ages.
chikei	darker lines (folds) in *ji*
choji-hamon	clove-shaped tempering line

Corsair	The Vought F4U Corsair was a US carrier-operated fighter-bomber during World War II. Its characteristic bent wings were striking.
efu (no) tachi	Efu tachi, also known as *hoso tachi*, were worn exclusively by high-ranking princes *(daimyo)* and the highest dignitaries at court. Characteristic features of the mounting *(koshirae)* are the *shitogi tsuba* and the *same* grip without handle winding. This mount serves primarily for ceremonial purposes and less for fighting. Efu tachi were made from the *Koto* to the *Showa* period.
fuchi	Handle clamp, front fitting part of the sword handle.
fukura	cutting edge at the tip of the sword
gendaito	Literally "modern sword"; term for a traditionally forged Japanese sword from the time after 1876 until today.
gimei	wrong signature
gunome-hamon	tempering line of a sequence of uniform short waves
gunome shoji	tempering line of a sequence of cloverleaf-shaped figures

gunto	Literally "army or military sword". The blade could be machine made, but also traditionally hand forged.
ha	cutting edge of the sword
habaki	blade collar
hada	"Skin", damask structure of the blade surface; visible forging pattern.
hada – ayasugi hada	horizontal and parallel running waves
hada – itame hada	like wood grain
hada – ko-itame hada	similar to *itame*, but smaller/denser grain
hada – konuka hada	Extremely dense *ko-mokume hada* with fine, homogeneous surface structure; characteristic of swords forged by the smiths of the Tadayoshi school in Hizen province.
hada – masame hada	horizontal and parallel running straight lines
hada – mokume hada	like burl wood
hada – muji hada	damask structure (to the naked eye) practically no longer recognizable
ha-machi	step from the cutting edge to the tang
hamon	tempering pattern

Heian period	794 – 1184
Hellcat	The Grumman F6F Hellcat was a US-American carrier-based fighter of the World War II.
hiro-maki	typical handle winding on Kai-gunto
Hizen-to	Sword(s) from the province of Hizen
horimono	engraving on the sword blade
iaido	"Way of drawing the sword", one of the Budo disciplines. The aim is to use the sword as weapon while drawing it.
ji	blade surface between *hamon* and *shinogi*
ji-tetsu	Appearance/color/texture of the blade surface, caused by the iron *(tetsu)* used in sword making.
kabuto-gane	sword pommel at *tachi*
kaishakunin	Selected second whose duty in *seppuku* is to behead the one who commits *seppuku* before the death struggle begins.
Kamakura period	1185 – 1333
kashira	pommel, rear fitting part of the sword handle

| katana | Sword blade with a tang hole and longer than two *shaku*. The katana is carried in the belt with the cutting edge facing upwards. The signature *(mei)* is located on the side *(ura)* of the tang *(nakago)* facing the wearer and is called *katana-mei*. |

katana-kaji — swordsmith

katana-mei — Signed as *katana*. When worn, the signature *(mei)* is located on the side *(ura)* of the tang *(nakago)* facing the wearer.

Kawasaki Ki-45 — The Kawasaki Ki-45 "Toryu" ("Dragon Slayer") was a Japanese twin-engine, two-seat fighter aircraft in World War II.

kawasatetsu — Iron sand extracted from rivers for the production of swords.

kiku — Chrysanthemum

kiku-mon — Imperial Chrysanthemum Crest

kiku-sui mon — Crest *(mon)* of the Minatogawa shrine. It depicts in a stylized form a chrysanthemum blossom on the waves of the Minatogawa River.

kiri-komi — battle scars

kissaki — point of the sword

kissaki – chu-kissaki	medium length sword point
kissaki – ko-kissaki	short sword point
kissaki – o-kissaki	long sword point
koshirae	sword mounting
koshirae – Gunto	army style mounting
koshirae – Kai-gunto	navy style mounting
koshi-zori	Sword blade with the curvature near the tang, also known as *bizen-zori*.
koto	„Old sword", production time 782 - 1596.
kuge	court nobility
kuri-jiri	tang end rounded
mantetsu	Manchurian steel; literally "iron *(tetsu)* from Manchuria".
mantetsu-to	sword made of Manchurian steel
mei	signature of the swordsmith
mei – gimei	wrong signature
mei – tachi-mei	As tachi signed sword. When worn, the signature *(mei)* is on the side facing away from the wearer *(omote)*.

mei – mumei	without signature
Meiji-Periode	Reign of Emperor Mutsuhito (Meiji-Tenno) from January 25, 1868 to July 30, 1912.
Meiji-Restauration	Disempowerment of the military government *(bakufu)* by Emperor Mutsuhito and renewal of the imperial power, establishment of a new political system according to western orientation, abolition of the corporative society, prohibition of sword carrying for samurai.
mekugi	bamboo peg for fixing the tang *(nakago)* in the handle *(tsuka)*
mekugi-ana	tang hole to hold the *mekugi*
menuki	Ornamental piece on the sword hilt, usually under the hilt winding; in the case of court mountings superficially attached to the *same*.
midareba	irregular tempering line
midare komi	irregular tempering line which continues into the *boshi*
Minatogawa mon	Crest *(mon)* of the Minatogawa shrine. It depicts in a stylized form a chrysanthemum blossom on the waves of the Minatogawa River.

Minatogawa shrine	Shinto shrine in Kobe, Japan. It was built to worship Kusunoki Masashige, who in 1336, after losing a battle on the Minatogawa River, committed Sepukku. His wife is shrined in a side shrine. At the Shrine, swords were traditionally forged for Japanese naval officers by renowned swordsmiths during World War II.
mon	crest, coat of arms
morohineri-maki	typical handle winding on *gunto*
moto-haba	width of the blade between *mune-machi* and *ha-machi*
moto-kasane	thickness of the blade at the *mune-machi*
mumei	without signature, unsigned
mune	back of the blade
mune – hira oder kaku ~	back of the blade flat
mune – ihori oder iori ~	back of the blade roof-shaped
mune – maru oder sono ~	back of the blade rounded
mune-machi	step from blade back to tang

Muromachi-jidai	Muromachi period, 1336 – 1573
nagasa	length of the blade from the tip of the sword to the *mune-machi*
nakago	tang of the sword
nakago – ha-agari kuri ~	tip of the tang asymmetrical U-shaped
nakago – kata-yamagata ~	tip of the tang asymmetrical V-shaped
nakago – kengyo ~	tip of the tang V-shaped
nakago – kiji-momo ~	tip of the tang trapezoidal, "like a kimono sleeve"
nakago – kiri oder kaku-ichi-monji ~	tip of the tang rectangular
nakago – kuri-jiri oder hira-yamagata ~	tip of the tang evenly rounded
nakago-jiri	tip of the tang
namban-tetsu	imported steel, literally "iron from the southern barbarians"
nanako	tiny dimpling like fish roe
nidai	second generation
nie	Due to the tempering process, martensite crystals lie on the surface of

the blade, which become visible through polishing and make the tempering line *(hamon)* on the blade surface appear like a band of countless dots strung together. One speaks of *nie* when the grain is still visible to the naked eye.

nihonto	japanese sword
Nihon Bijutsu Token Hozon Kyokai	"Society for the Preservation of the Japanese Art Sword"
Nihonto Tanren Kai	"Japanese Swordsmith Center" at *Yasukuni Shrine*
Nihonto Token Hozon Kai	"Society for the Preservation of the Japanese Sword"
nioi	Due to the tempering process, martensite crystals lie on the surface of the blade, which become visible through polishing and make the tempering line *(hamon)* on the blade surface appear like a silver band. One speaks of *nioi* when the individual martensite crystals are so fine and so close together that the grain is hardly visible to the naked eye.
notare-hamon	undulating tempering line
obi	belt

omote	The side of the tang facing away from the carrier when carrying a sword.
origami	Expertise, certificate of authenticity; colloquially also "paper" or "papers".
oshigata	Abrasion of the sword tang on thin paper, comparable to a blueprint of the sword tang. Sometimes *oshigata* of the whole blade including the tang are made.
oyioroi	great armor
PFC	Private First Class
rikugun jumei tosho	Swordsmith certified by the Imperial Japanese Army and working on behalf of the army.
saki-haba	height/width of the blade at the *yokote*
saki-kasana	thickness of the blade at the *yokote*
same	stingray skin
sandai	third generation
saya	scabbard of the sword
Seki	city in the Japanese prefecture of Gifu

Sengoku-jidai	Sengoku period, 1477 - 1573; entered the history of Japan as the "Time of the Fighting Empires" or the "Time of the Warring Countries"
sepukku	Ritualized suicide practiced in Japan by the Samurai to wipe out shame and restore family honor after losing face.
shaku	Japanese unit of measurement; 1 *shaku* = 30,3 cm.
shinogi	horizontal dividing line between *ji* and *shinogi-ji*
shinogi-ji	blade surface between *shinogi* and *mune*
shinogi-zukuri or hon-zukuri	Curved blade with horizontal dividing line between *ji* and *shinogi-ji* and vertical dividing line between blade and tip of the sword.
shinsa	Official appraisal of a blade by a committee of sword experts, e.g. the NBTHK or the NTHK.
shinsaku-to	new forged sword
shinto	„New sword", production time 1597 - 1780.
shin-shinto	"New-new sword", production time 1781 - 1876.

shirasaya	Literally "white sheath", made of magnolia wood. In contrast to the *koshirae*, in which the blade is carried, the shirasaya is used to store the blade.
shitogi-tsuba	Crosswise to the blade aligned guard, similar to European guards, but wider and more magnificently designed.
shodai	first generation
Shogun	leader of the warrior nobility
shogunat	administrative apparatus of the Shogun
shokonsha	Shinto shrine for the worship of war dead
Showa-Periode	Reign of Emperor Hirohito (Showa-Tenno) from December 25, 1926 to January 7, 1989. "The first half of the Showa period was the heyday of Japanese imperialism. Under the influence of various generals, the young emperor ordered the Japanese army to attack China, the Philippines and the United States. After dropping the atomic bombs on Hiroshima and Nagasaki, Emperor Hirohito was the first Japanese emperor to speak to his people personally. In a radio message, he announced Japan's surrender and at the same time renounced his

claim to divinity to clear the way for a new peaceful and democratic social order. Due to the high esteem in which the emperor was held, the abolition of the monarchy was not considered; although the Tenno lost almost all political rights of participation in the new constitution, he remained highly respected in his remaining 44 years of government." (Source Wikipedia)

sori	Depth of blade curvature, other spelling *zori*.
Stuka	German abbr. for "Sturzkampfbomber", English dive bomber.
sugata	blade shape
sugu-ba hamon	Straight tempering line; also *suguha hamon*.
suguha-hamon	straight tempering line
sun	Japanese unit of measurement; 1 *sun* = 3,03 cm.
suriage nakago	shortened sword tang
tachi	Sword blade with a tang hole and longer than two *shaku*. In contrast to the *katana*, the *tachi* is carried hanging with the edge down. The signature is located on the side of the tang

	that is turned away from the wearer *(omote)* and is called *tachi-mei*.
tachi-mei	As *tachi* signed sword. When worn, the signature *(mei)* is located on the side *(omote)* of the tang *(nakago)* facing away from the wearer.
Taisho period	Reign of Emperor Yoshihito (Taisho-Tenno) from July 30, 1912 to December 25, 1926.
tamahagane	Literally "jewel steel"; raw steel obtained by a traditional Japanese smelting process for blade production.
Tanegashima rifle	When a Chinese ship stranded at Cape Kadokura in the south of Tanegashima Island in 1543, there were also Portuguese on board, carrying archebus. The Daymio of the island, Tanegashima Tokitaka, bought one or two rifles from the Portuguese for a large sum of money. Within a short period of time, 20,000 copies of the weapon were produced, which revolutionized warfare in Japan.
tanto	Sword blade with a tang hole and a length under a *shaku*.
tatara	Traditional Japanese furnace for the extraction of *tamahagane*.

tetsu	iron
tocho	sword knot
tsuba	sword guard
tsuka	sword hilt
tsuka-ito	silk or cotton tape for handle winding
ubu-ha	Ricasso; non sharpened part of the blade.
ubu nakago	unshortened tang
uchigatana	In the Muromachi period, it became a custom to carry a second sword in the belt *(obi)* in addition to the *tachi*. The *uchigatana ("cutting sword")* had a blade about 60 cm long. Unlike the *tachi*, which was designed for fighting on horseback, the *uchigatana* was used in close combat.
ura	The side of the tang facing the carrier when carrying a sword.
utsuri	Milky white appearance in *ji*, sometimes also called reflection of *hamon* in *ji*.
wakizashi	Sword blade with a tang hole and a length between one and two *shaku*.

western steel	General for imported steel from Europe.
yamasatetsu	Iron sand extracted from the mountains.
Yasukuni shrine	Shinto shrine in Tokyo, Japan. Here all fallen hero souls are remembered who lost their lives during and since the *Meiji Restoration* on the side of the Imperial armies. Originally founded in 1869 as *Tokyo Shokonsha (Tokyo Shrine for Summoning the Spirits of the Dead)* in the Chiyoda district north of the Imperial Palace, the Tenno raised it to the rank of *Bekkaku Kanpeisha (Imperial Shrine of the Special Class)* in 1879 and gave it the name *Yasukuni-jinja (Shrine of Peaceful Land)*. In the period from 1933 to 1945, a total of 8,100 swords were forged at the shrine. They fulfill all the criteria for the Japanese sword as a work of art and are highly sought after and coveted in collector circles.
Yasukuni-to	Sword forged at *Yasukuni Shrine*.
yasurime	file marks on the tang
yamasatetsu	Iron sand mined in the mountains for the production of swords.

yo	separate fine sections/spots of *nie* or *nioi* between hardening line and cutting edge
yokote	vertical dividing line between blade and tip of the sword
Zero	The Mitsubishi.A6M, also known as the "Zero", was a carrier-based Japanese fighter in World War II. The designation Zero comes from the last digit of the year of entry into service, 2600 Japanese calendar (1940).
zori	Depth of blade curvature; different spelling *sori*.

List of References

Alperovitz, Gar
Atomic Diplomacy: Hiroshima and Potsdam,
Vintage Books 1965

Couch, Paul and **Matsuoka, Yumiko**
"Thoughts on Nihonto – Gendaito",
ISF-AL/GA Newsletter, Mai 2002, Vol. 3, Issue 3

Dawson, Jim
"Swords of Imperial Japan, 1868 – 1945",
Stenger-Scott Publishing, 1996

Fujishiro, Yoshio und **Matsuo**
Nihon Toko Jiten Shinto-hen
Tokyo, 1971, 5. Auflage

Fuller, Richard and Gregory, Ron
Military Swords of Japan 1868 – 1945
Arms and Armour Press, London - New York – Sydney, 1986

Gewitsch, Michael
„Die sieben Tugenden der Samurai",
Ausarbeitung zum 1. DAN Jiujitsu, 2013

Goepper, Roger, Oishi Shinzaburo, Tokugawa Yoshinobu
Shogun, Kunstschätze und Lebensstil eines japanischen Fürs-
ten der Shogun-Zeit
Katalog zur Ausstellung Haus der Kunst München,
Toppan Printing Co., Ltd., Tokyo, September 1984

Hagenbusch, Michael
Die Kunst der Samurai
Katalog zum Ersten Europäischen Symposium, herausgegeben von Trudel Klefisch, Köln im August 1984

Han, Bing Siong
Japanese Swords in Dutch Collections
De Nederlandse Token Vereniging, 2003

Hill, P. Joshua, Professor Koshiro,Yukiko
Remembering the Atomic Bomb,
FreshWriting 15 December 1997

Icke-Schwalbe, Lydia
Das Schwert des Samurai,
Brandenburgisches Verlagshaus, 3. Auflage Berlin 1990

Kapp, Leon and **Hiroko, Yoshindo Yoshihara**
The Craft of the Japanese Sword
Kodansha International, Japan, 1. Auflage 1987

Kazuo Tokuno
TOKO TAIKAN
Tokyo, Kogai Shuppan, 2004

Kensho Furuya
Proper Use of Swords
Aikido Center of Los Angeles
The Aiki Dojo Newsletter, November 2020
http://www.aikidocenterla.com

Kishida, Tom
The Yasukuni Swords, Rare Weapons of Japan
1933 – 1945
Kodansha Europe Ltd., 1. Auflage 2004

Kokan Nagayama
The Connoisseurs Book of Japanese Swords
Kodansha International, Japan, 1. Auflage 1997

Kümmel, Otto
Das Kunstgewerbe in Japan
Schmidt & Co. Berlin, 3. Auflage 1922

Mauer, Kuno
Die Samurai
Econ Verlag GmbH, Düsseldorf, 1. Auflage 1981

Nickel, Helmut
"The Japanese Blade: Technology and Manufacture"
http://www.metmuseum.org/toah/hd/japb/hd_japb.htm

Obata Toshishiro
Crimson Steel: The Sword Technique of the Samurai
Dragon Books, Oktober 1987

Obata Toshishiro
"Swords and Tradition"
https://kenjutsu-ryu.livejournal.com/29096.html

Ohnuki-Tierney, Emiko
Kamikaze, Cherry Blossoms and Nationalisms
The University of Chicago Press, 12. November 2002

Sachse, Manfred
Damaszener Stahl
Verlag Stahleisen, Düsseldorf, 2. erweiterte Auflage 1993

Sato Kanzan
The Japanese Sword
Kodansha International and Shibundo, Japan, 1983

Slough, John Scott
Modern Japanese Swordsmiths 1868 – 1945
Rivanna River Company, 1. Auflage 2001

Sly, Christopher
"More Thoughts On Gendaito", Sept. 1992, updated
by Bowen, Chris and Massey, Denny, March 2015
http://www.nihontocraft.com/Gendaito.html

South Manchuria Railway Co.Ltd. Dalian Railway Factory Sword Works
"Kōa Issin"
Published by "The South Manchuria Railway Company", July 25, 1939

Stein, Richard
Japanese Sword Guide, Minamoto Yoshichika
http://www.japaneswordindex.com/yoshchik.htm

Tillman, Barrett
Carrier Battle in the Philippine Sea: The Marianas Turkey Shoot
Specialty Press, 1994

Ugaki Matome
Fading Victory: The Diary of Admiral Matome Ugaki, 1941-1945
University of Pittsburgh Press, 1991

Yamamoto Tsunetomo
Hagakure, der Weg des Samurai
Piper Verlag GmbH München, 5. Auflage 2011

Yumoto, John M.
Das Samuraischwert, Ein Handbuch
Ordonanz-Verlag Strebel GmbH, Wiesbaden, 2004

....

Accompanying Literature
Other Authors Mentioned in this Book

Engermann, Andreas
Einen Bessern findst du nicht
© Kindler-Verlag München,
Lizenzausgabe des Verlages Buch und Welt, Klagenfurt

Köppen, Edlef
Heeresbericht
Nikol Verlagsgesellschaft mbH & Co. KG, Hamburg, 2012

Remarque, Erich Maria
Im Westen nichts Neues
Kiepenheuer & Witsch, Köln, 3. Auflage 2011

Remarque, Erich Maria
Zeit zu leben und Zeit zu sterben
Kiepenheuer & Witsch, Köln, 1. Auflage 2018

WIKIPEDIA, sources in alphabetical order

https://en.wikipedia.org/wiki/28_cm_howitzer_L/10

https://de.wikipedia.org/wiki/Atombombenabwürfe_auf_Hiros
hima_und_Nagasaki#Gegner_der_Abwürfe

https://en.wikipedia.org/wiki/Attack_on_Pearl_Harbor

https://en.wikipedia.org/wiki/Battle_of_Kokoda

https://en.wikipedia.org/wiki/Battle_of_Singapore

https://de.wikipedia.org/wiki/Belagerung_von_Port_Arthur

https://de.wikipedia.org/wiki/Bougainville

https://en.wikipedia.org/wiki/Bougainville_Campaign

http://en.wikipedia.org/wiki/Dai_Nippon_Butoku_Kai

http://de.wikipedia.org/wiki/Europäische_Expansion

http://de.wikipedia.org/wiki/Gunto

https://de.wikipedia.org/wiki/Hávamál

https://en.wikipedia.org/wiki/Japanese_invasion_of_Malaya

http://en.wikipedia.org/wiki/Japanese_sword

https://de.wikipedia.org/wiki/Kohima

http://de.wikipedia.org/wiki/Meijin

http://de.wikipedia.org/wiki/Minatogawa-Schrein

https://de.wikipedia.org/wiki/Mitsubishi_Ki-21

http://de.wikipedia.org/wiki/Nakayama_Hakudo

http://en.wikipedia.org/wiki/Nakayama_Hakudo

http://de.wikipedia.org/wiki/Ninigi

http://de.wikipedia.org/wiki/Nogi_Maresuke

http://de.wikipedia.org/wiki/Otto_Kümmel

http://de.wikipedia.org/wiki/ Ōnishi_Takijirō

https://en.wikipedia.org/wiki/Takijirō_Ōnishi

http://de.wikipedia.org/wiki/Pazifikkrieg

http://en.wikipedia.org/wiki/Privy_Council_Japan

https://de.wikipedia.org/wiki/Schlacht_am_Minatogawa

https://de.wikipedia.org/wiki/Schlacht_um_Manila_(1945)

https://de.wikipedia.org/wiki/Schlacht_um_Nanchang

https://de.wikipedia.org/wiki/Shimpū_Tokkōtai

http://de.wikipedia.org/wiki/Sugiyama_Hajime

http://en.wikipedia.org/wiki/Takeji_Nara

http://en.wikipedia.org/wiki/Tatara_(furnace)

http://de.wikipedia.org/wiki/Todesmarsch_von_Bataan

https://en.wikipedia.org/wiki/Tomoyuki_Yamashita

http://de.wikipedia.org/wiki/Ugaki_Matome

http://de.wikipedia.org/wiki/Ulfberht

https://de.wikipedia.org/wiki/USS_Arizona_(BB-39)

https://en.wikipedia.org/wiki/USS_Belleau_Wood_(CVL-24)

https://de.wikipedia.org/wiki/USS_Bunker_Hill_(CV-17)

http://de.wikipedia.org/wiki/USS_Enterprise_CV-6

http://de.wikipedia.org/wiki/USS_Hornet_(CV-8)

https://en.wikipedia.org/wiki/USS_St._Lo

https://en.wikipedia.org/wiki/USS_St._Louis_(CL-49)

https://de.wikipedia.org/wiki/USS_Louisville_(CA-28)

https://de.wikipedia.org/wiki/USS_Missouri_(BB-63)

https://de.wikipedia.org/wiki/Yamashita_Tomoyuki

https://de.wikipedia.org/wiki/Yamato_(Schiff,_1941)

http://de.wikipedia.org/wiki/Yasukuni-Schrein

http://en.wikipedia.org/wiki/Yukio_Seki

http://de.wikipedia.org/wiki/Zwischenfall_an_der_Marco-Polo-Brücke

WWW, further internet sources

https://www.aoijapan.com/

https://www.aoijapan.net/katana-minamoto-yoshichika-saku-kore/

http://home.earthlink.net/~steinrl/
Japanese Sword Guide by Richard Stein

http://www.japaneseswordindex.com/
Japanese Sword Guide by Richard Stein

https://www.britannica.com/biography/Yamashita-Tomoyuki

http://www.bushido.de/philsophie.htm

http://www.chiran-tokkou.jp/learn/pilots/FujiiHajime.html

https://www.historynet.com/general-tomoyuki-yamashita.htm

https:// www.cicero.de/aussenpolitik/atombombe-auf-nagasaki-japan-haette-auch-ohne-bombe-kapituliert/59654
Scherer, Klaus, Interview mit Martin Sherwin, 3. August 2015

http://www.kamikazeimages.net/stories/lcsl114/index.htm

http://www.metmuseum.org/toah/hd/japb/hd_japb.htm

http://www.nihonto.com.au/html/minamoto_yoshichika_tachi.html, JSS Newsletter, auszugsweise veröffentlicht.

https://www.nihonto.com/the-yasutsugu-school

https://www.poetryfoundation.org/poets/takijiro-onishi

http://www.samuraisam.net/tachiofyoshichika.html

https://www.spiegel.de/spiegel/print/d-13525110.html
„ General Yamashita und sein verfluchtes Gold",
DER SPIEGEL 26/1987

https://de.statista.com/statistik/daten/studie/1086264/umfrage/g
eschaetzte-zivile-todesopfer-und-verletzte-in-hiroshima-und-
nagasaki/

https://teachingamericanhistory.org/library/document/potsdam-
declaration/

https://ww2db.com/person_bio.php?person_id=297
World War II Database, Home – People – Yukio Seki

http://www.ussarizona.org/stories/uss-arizona-survivor-
stories/107-lancaster-james-daniel-usn

https://www.warhistoryonline.com/world-war-ii/the-tragic-tale-
of-hajime-fujii-a-kamikaze-fighter.html

http://www.worthpoint.com/worthopedia/rare-mint-antique-
japanese-samurai-katana-sword

https://www.zeit.de/online/2009/35/atombombe-hiroshima,
Knauß, Ferdinand, Ein Experiment mit 70.000 Toten